IJPHM

International Journal of Prognostics and Health Management

The International Journal of Prognostics and Health Management (IJPHM) is the premier online journal related to multidisciplinary research on Prognostics, Diagnostics, and System Health Management. IJPHM is the archival journal of the Prognostics and Health Management (PHM) Society. It exists to serve the following objectives:

- To provide a focal point for dissemination of peer-reviewed PHM knowledge.
- To promote multidisciplinary collaboration in PHM education and research.
- To encourage and assure establishment of professional standards for the practice of PHM.
- To improve the professional and academic standing of all those engaged in the practice of PHM.
- To encourage governmental and industrial support for research and educational programs that will improve the PHM process and practice.

The Journal supports these goals by providing a venue for archival publication of peer-reviewed results from research and development in the area of PHM. We define PHM as a system engineering discipline focused on assessing the current status and well as predicting the future condition of a component and/or system of components. PHM is broader than any single field of engineering: it draws from electrical, electronics, mechanical, civil, and chemical engineering, computer and materials science, reliability, test and measurement, artificial intelligence, physics, and economics. IJPHM seeks to publish multidisciplinary articles from industry, academia, and government in diverse application areas such as energy, aerospace, transportation, automotive, and industrial automation. IJPHM is dedicated to all aspects of PHM: technical, management, economic, and social.

IJPHM

International Journal of Prognostics and Health Management

2011   **Vol. 2 Issues 1-2**

Table of Contents

http://phmsociety.org
Free and open access to full text papers worldwide.

Uncertainty Quantification in Fatigue Crack Growth Prognosis

Shankar Sankararaman[1], You Ling[2], Christopher Shantz[3], and Sankaran Mahadevan[4]

[1,2,3,4] *Department of Civil and Environmental Engineering, Vanderbilt University, Nashville, TN-37235, USA.*

shankar.sankararaman@vanderbilt.edu
you.ling@vanderbilt.edu
chris.shantz@vanderbilt.edu
sankaran.mahadevan@vanderbilt.edu

ABSTRACT

This paper presents a methodology to quantify the uncertainty in fatigue crack growth prognosis, applied to structures with complicated geometry and subjected to variable amplitude multi-axial loading. Finite element analysis is used to address the complicated geometry and calculate the stress intensity factors. Multi-modal stress intensity factors due to multi-axial loading are combined to calculate an equivalent stress intensity factor using a characteristic plane approach. Crack growth under variable amplitude loading is modeled using a modified Paris law that includes retardation effects. During cycle-by-cycle integration of the crack growth law, a Gaussian process surrogate model is used to replace the expensive finite element analysis. The effect of different types of uncertainty – physical variability, data uncertainty and modeling errors – on crack growth prediction is investigated. The various sources of uncertainty include, but not limited to, variability in loading conditions, material parameters, experimental data, model uncertainty, etc. Three different types of modeling errors – crack growth model error, discretization error and surrogate model error – are included in analysis. The different types of uncertainty are incorporated into the crack growth prediction methodology to predict the probability distribution of crack size as a function of number of load cycles. The proposed method is illustrated using an application problem, surface cracking in a cylindrical structure

1. INTRODUCTION

The scientific community has increasingly resorted to the use of computational models to predict the performance of

engineering components and systems so as to facilitate risk assessment and management, inspection and maintenance scheduling, and operational decision-making. Model-based prognosis, i.e. predicting the performance of a system using a physics-based model is promising for health management. However, no model can perfectly represent the system and hence it is necessary to include model form errors and model uncertainty in the prognosis. Secondly, complex engineering systems may have to be modeled using multiple models that interact with one another. In such cases, each model has its own sources of error/uncertainty and the interaction between the errors of multiple models is non-trivial. Some errors are deterministic while some others are stochastic. Systematic methods are needed to quantify the uncertainty and confidence associated with the model prediction. Hence, prognosis methods need the following capabilities: (1) integration of multiple models, (2) quantification of different types of uncertainty and error (physical variability, data uncertainty, and model uncertainty), and (3) integration of the various types of uncertainty to calculate the overall uncertainty in the results of prognosis.

This paper develops an uncertainty quantification methodology to meet the above needs, and uses the problem of fatigue crack growth to illustrate such development. The objective of this problem is to predict the crack growth in a structural component as a function of number of load cycles. Mechanical components in engineering systems are often subjected to cyclic loads leading to fatigue, crack initiation and progressive crack growth. It is essential to predict the performance of such components to facilitate risk assessment and management, inspection and maintenance scheduling and operational decision-making. Researchers have pursued two different kinds of methodologies for fatigue life prediction. The first method is based on material testing (to generate S-N, ε –N curves) and use of an assumed damage accumulation rule. In this method, specimens are subjected to repeated cyclic loads under laboratory conditions. Hence the results are specific to the geometry of the structure as well as the nature of loading. Further, the performance of these components under field conditions is significantly different from laboratory observation, due to various sources of uncertainty accumulating in the field that render experimental studies less useful. Hence, this methodology cannot be used directly to predict the fatigue life of practical applications wherein complicated structures subjected to multi-axial loading.

The second method for fatigue life prediction is based on principles of fracture mechanics and crack growth analysis. A crack growth law is assumed and the progressive growth of the crack is modeled. However, this is not straightforward. Fatigue crack growth is a stochastic process and there are different types of uncertainty – physical variability, data uncertainty and modeling errors, associated with it. Uncertainty appears at different stages of analysis and the interaction between these sources of uncertainty cannot be modeled easily. Further, the application of crack growth principles to complicated structures, subjected to multi-axial variable amplitude loading requires repeated evaluation of finite element analysis which makes the computation expensive.

Some of these problems have been investigated by researchers in detail. The first problem in using a crack growth model is that the initial crack size is not known. This issue is further complicated by the fact that small crack growth propagation is anomalous in nature. This problem was addressed by the introduction of an equivalent initial flaw size (EIFS) nearly thirty years ago. The concept of EIFS was introduced to by-pass small crack growth analysis and to substitute an initial crack size in long crack growth models such as Paris' law. However EIFS does not represent any physical quantity and cannot be measured using experiments. Initially, certain researchers used empirical crack lengths between 0.25 mm and 1 mm for metals (JSSG, 1998; Gallagher et al., 1984; Merati et al., 2007). Later, several researchers (Yang, 1980; Moreira et al., 2000; Fawaz, 2000; White et al., 2005; Molent et al., 2006) used back-extrapolation techniques to estimate the value for equivalent initial flaw size. Recently, Liu and Mahadevan (2008) proposed a methodology based on the Kitagawa-Takahashi diagram (Kitagawa and Takahashi, 1976) and the El-Haddad Model (Haddad et al., 1979) to derive an analytical expression for the equivalent initial flaw size. The current research work uses this concept to calculate the statistics of EIFS from material properties such as threshold stress intensity factor and fatigue limit. These material properties are calculated from experimental data and the associated data uncertainty due to measurement errors, sparseness of data, etc. needs to be taken into account.

The next step in fatigue crack growth prognosis is to choose a crack growth model. There are many crack growth models available in literature. In this paper, a modified Paris law is used as the crack growth law for the sake of illustration, but an error term (treated as a random variable) is added to represent the fitting error since experimental data were used to estimate the coefficients of the Paris model. Further, the model coefficients are also treated as random variables. The effects of variable amplitude loading are considered by including retardation effects along with modified Paris' law. Several models (Wheeler, 1972; Schjive, 1976; Noroozi et al., 2008) have been proposed to tackle variable amplitude loading conditions and this paper uses Wheeler's retardation model (Wheeler, 1972) only for illustration purposes. Further, only coplanar cracks have been considered for analysis.

The modified Paris law based on linear elastic fracture mechanics calculates the increase in crack size as a function of the stress intensity factor, during each loading cycle. The stress intensity factor, in turn, is a function of the current crack size, crack configuration, geometry of the structural component and loading conditions. If structures with complicated geometry are subjected to multi-axial loading, then the stress intensity factor needs to be calculated through expensive finite element analysis, at every loading cycle. This paper replaces the finite element analysis with a surrogate model, known as the Gaussian process (GP) interpolation. Several finite element analysis runs are used to train this surrogate model and then, the surrogate model is used to predict the stress intensity factor, to be used in the crack growth law. There are two types of errors in this procedure. First, the finite element analysis has discretization error that needs to be accounted for while training the surrogate model. Second, the surrogate model adds further uncertainty since it is obtained by fitting the model to the (finite element) training data.

In addition to the above mentioned model uncertainty and data uncertainty (used to calculate the EIFS), natural variability in many input variables introduces uncertainty in model output. The loading on the structure is usually random in nature. A variable amplitude multi-axial loading history consisting of bending and torsion is illustrated in this paper. Natural variability also includes variability in material properties, geometry and boundary conditions. The variability in certain material properties such as fatigue limit and threshold stress intensity factor is considered while deriving the statistical distribution of EIFS. The geometry of the specimen and boundary conditions are considered deterministic in this research work.

The main focus of this paper is to investigate in detail each source of uncertainty and propose a methodology that can effectively account for all of them. Finally, the developed framework is used to predict the probabilistic fatigue life of the structure.

The next section reviews the existing literature on this topic and motivates the current study. Section 3 presents the algorithm used in this paper to predict the fatigue life of structures with complicated geometry and subjected to variable amplitude, multi-axial loading. The various sources of uncertainty in this procedure are discussed in Section 4. Section 5 presents the proposed framework for uncertainty quantification in crack growth prediction. Section 6 illustrates the methodology through an example, considering cracking in a cylindrical structure.

2. LITERATURE REVIEW

Numerous studies have dealt with methods for uncertainty quantification in prognosis. Most of these studies have focused mainly on natural variability; sources of data uncertainty and model uncertainty have not been considered in detail. Hemez (2005) discusses uncertainty quantification in prognosis and uncertainty propagation techniques that deal only with physical variability. Chelidze and Cusumano (2004) demonstrate a dynamical systems approach to prognosis in an electromechanical system and estimate the uncertainty in the prognosis results. Saha and Goebel (2007) discuss diagnostics and prognostics of batteries using Bayesian techniques. Medjaher et al (2009) use dynamic Bayes networks for prognosis of industrial systems. These studies mostly consider physical variability; modeling errors and their sources are not analyzed in detail; and data uncertainty due to sparse data is not considered.

The "damage prognosis" project at Los Alamos national laboratory (Doebling and Hemez, 2001; Hemez et al., 2003; Farrar et al., 2004; Farrar and Lieven, 2006) considered the problem of fatigue cracking in detail and proposed sampling techniques to predict crack growth in composite plates; the error between prediction and observation was also characterized. Loading (uniaxial impact loading) conditions and geometric and material properties are treated as random variables. Surrogate models were used to replace expensive finite element models, and included in a sampling based framework for uncertainty propagation. Finite element analysis results were used to train the surrogate models, but the discretization error was not quantified. Further, the errors due to usage of surrogate models, errors in crack growth model, etc. were not addressed.

Besterfield et al. (1991) combined probabilistic finite element analysis with reliability analysis to predict crack growth in plates. Random mixed mode loading cycles, physical variability in material properties, randomness in crack configuration (size, position and angle) were considered. However, the implementation of probabilistic finite element analysis is computationally expensive for structures with complicated geometry. Other sources of uncertainty such as data uncertainty and model uncertainty were not considered.

Patrick et al (2007) introduced an online fault diagnosis and failure prognosis methodology applied to a helicopter transmission component. A crack growth model (Paris law) was used for fatigue life prediction. Bayesian techniques were implemented to infer the initial crack size, which was used for probabilistic fatigue life prediction using particle filter techniques. Other sources of uncertainty such as error in Paris law, variability in model parameters, and randomness in loading were not considered.

Gupta and Ray (2007) developed algorithms for online fatigue life estimation that relied on time series data analysis of ultrasonic signals and were built on the principles of symbolic dynamics, information theory and statistical pattern recognition. Physical variability in material geometry (surface defects, voids, inclusions, sub-surface defects), minor fluctuations in environmental conditions and operating conditions were used to quantify the uncertainty in detection which was further used to quantify the uncertainty in prognosis.

Pierce et al (2007) discussed the application of interval set techniques to the quantification of uncertainty in a neural network regression model of fatigue life, applied to glass fiber composite sandwich materials. This paper only considered the uncertainty in input data and other sources of uncertainty were not investigated in detail.

Orchard et al (2008) used the method of particle filters for uncertainty management in fatigue prediction. However, the various sources of uncertainty were not clearly delineated and considered in the analysis. While the use of conditional probability has been recommended for probabilistic predictions, this turns out to be expensive when variable amplitude loading cycles are considered, as the ensemble of predictions grows in size as a function of the number of loading cycles.

Papazian et al (2009) developed a structural integrity prognosis system (SIPS), based on collaboration between sensor systems and advanced reasoning methods for data fusion and signal interpretation, and modeling and simulation. Probabilistic principles such as likelihood and conditional probability were used to compare model predictions and sensor data. While measurement errors and sensor data were considered in detail, solution errors, variability of model parameters, randomness in loading, etc were not considered.

Thus, past studies on uncertainty quantification in prognosis have ignored several sources of uncertainty or not investigated them in detail. Physical variability (such as randomness in loading conditions, material properties, etc.) has been mainly studied by researchers, whereas other sources of uncertainty such as data uncertainty and model uncertainty have not been fully addressed.

This paper proposes a framework which can effectively account for different sources of uncertainty – physical variability, data uncertainty, and model uncertainty.

Data uncertainty arises due to the use of sparse data to construct probability distributions for input parameters. In previous studies, the input parameters are usually assumed to have completely known distributions (usually, normal) and prognosis is carried out using these distributions. In some cases, there may be enough data available to quantify such precise distributions. However, in many cases, it is impossible to construct precise probability distributions using a few available data (sparse data). Hence, there is uncertainty in the probability distributions constructed from such data. Therefore, this paper proposes a methodology where (1) the uncertainty in the distribution parameters is quantified using a resampling technique, and (2) an overall unconditional probability distribution of the quantity of interest, that includes the contribution of data uncertainty, is calculated.

When model-based methods are used for prognosis, it is essential to account for the different types of model uncertainty and errors. The significance of model uncertainty increases when there are multiple interacting models, because the quantification of the combined effect of the different sources of model uncertainty is non-trivial. In this paper, the algorithm for crack growth propagation uses multiple interacting models – finite element model, surrogate model, crack growth law, etc. First, the paper proposes methods to quantify the uncertainty/error in each of the individual models. It is important to note that some errors are deterministic (finite element discretization error) while some others are stochastic (crack growth law uncertainty, surrogate model prediction uncertainty), and they occur at different stages of the analysis. Therefore, the quantification of the overall uncertainty due to the combination of multiple sources of model uncertainty/errors is not trivial. This paper proposes a methodology for overall uncertainty quantification, where deterministic errors are addressed by correcting and stochastic errors are addressed through sampling.

Thus the contributions of this paper are: (1) connect different models such as finite element model, surrogate model, crack growth law, etc. efficiently; (2) quantify the uncertainty in each model separately; (3) treat deterministic model errors and stochastic model errors separately; (4) include physical variability as well as data uncertainty; and (5) quantify the overall uncertainty in crack growth prediction by correcting deterministic errors and sampling stochastic errors. The major advantage of the proposed methodology is that it provides a framework for including not only physical variability, but also data uncertainty and multiple model errors (both deterministic and stochastic). The various sources of uncertainty are discussed in detail,

later in Section 4. Prior to that, the algorithm for crack growth propagation is outlined in the following section.

3. CRACK GROWTH PROPAGATION

Consider the growth of an elliptic crack. A schematic of the crack growth is shown in Figure 1.

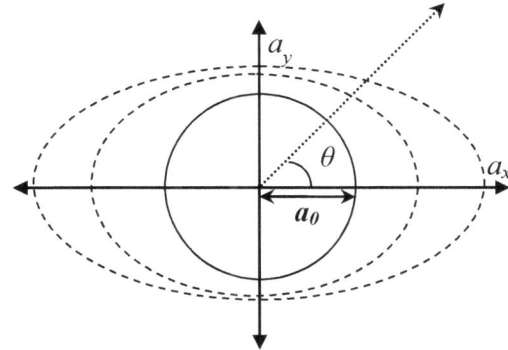

Figure 1. Elliptic Crack Growth

In Figure 1, a_x denotes the length of the semi-major axis and a_y denotes the length of the semi-minor axis. The aspect ratio, calculated as ratio between a_x and a_y is denoted by γ. If θ denote the angle of orientation, then a_x corresponds to $\theta = 0°$ and a_y corresponds to $\theta = 90°$. Crack growth laws such as Paris law (applicable to long cracks) predict the increase in crack size as a function of stress intensity factor, which in turn depends on the current crack size (a_x, a_y), aspect ratio (γ), angle of orientation (θ) and loading (L). In this paper, a has been used to denote the crack size in two directions, i.e. $a = [a_x, a_y]$. Hence, the two dimensional array a contains information about aspect ratio (β) as well. Starting with an initial crack size (a_0), the growth of the crack can be modeled and the crack size after a given number of cycles can be calculated. However, the initial crack size cannot be calculated exactly. The concept of EIFS was proposed to tackle this problem. Starting with the introduction of EIFS, this section explains the various steps involved in using a crack growth model to predict the crack size as a function of number of cycles.

3.1 Use of EIFS in Crack Growth Law

The rigorous approach to fatigue life prediction would be to perform crack growth analysis starting from the actual initial flaw, accounting for voids and non-metallic inclusions. If the initial crack size is large, then long crack growth models such as Paris' law can be used directly. However, this is not the case in most materials. Hence the long crack growth model cannot be used directly. A schematic plot of the long crack and short crack growth curves is given in Figure 2.

This paper uses a long crack model for fatigue crack growth analysis; the short crack growth calculations are bypassed through the use of an equivalent initial flaw size, as explained later in this section.

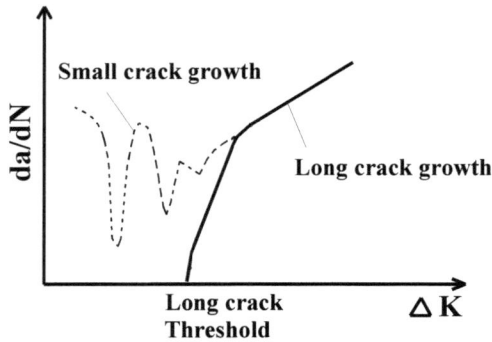

Figure 2. Schematic of Crack Growth

Consider any long crack growth law used to describe the relationship between da/dN and ΔK, where N represents the number of cycles, a represents the crack size and ΔK represents the stress intensity factor. This paper uses a modified Paris' law with Wheeler's retardation model as:

$$da/dN = \varphi^{r} C\,(\Delta K)^{n}(1-\Delta K_{th}/\Delta K)^{m} \qquad (1)$$

Note that several models (Wheeler, 1972; Schjive, 1976; Noroozi et al., 2008) have been proposed to tackle variable amplitude loading conditions. This paper uses a Wheeler's retardation model (Wheeler, 1972) only to illustrate the proposed uncertainty quantification methodology, and other appropriate models can also be used instead of the Wheeler model. In Eq. (1), φ^{r} refers to the retardation parameter (Sheu et. al., 1995), and is equal to unity if $a_i + r_{p,i} > a_{OL} + r_{p,OL}$ where a_{OL} is the crack length at which the overload is applied, a_i is the current crack length, $r_{p,OL}$ is the size of the plastic zone produced by the overload at a_{OL}, and $r_{p,i}$ is the size of the plastic zone produced at the current crack length a_i. Else, φ^{r} is calculated as shown in Eq. (2).

$$\varphi^{r} = (r_{p,i}\,/\,(a_{OL}+r_{p,OL}-a_{i}))^{\lambda} \qquad (2)$$

In Eq. (2), λ is the curve fitting parameter for the original Wheeler model termed the shaping exponent (Yuen et al., 2006). Song. et al. (2001) observed that crack growth retardation actually takes place within an effective plastic zone. Hence the size of the plastic zone can be calculated in terms of the applied stress intensity factor (K) and yield strength (σ) as:

$$r_p = \alpha\,(K/\sigma)^2 \qquad (3)$$

In Eq. (3), α is known as the effective plastic zone size constant which is calculated experimentally (Yuen et. al., 2006). The retardation model parameters are calibrated for

particular experimental conditions, which need to be matched to the problem at hand for proper application. The expressions in Eq. (2) and Eq. (3) can be combined with Eq. (1) and used to calculate the crack growth as a function of number of cycles. In each cycle, the stress intensity factor can be expressed as a function of the crack size (a), loading (L) and angle of orientation (θ). Hence, the crack growth law in Eq. (1) can be rewritten as:

$$da/dN = g(a,L,\theta) \qquad (4)$$

The concept of an equivalent initial flaw size was proposed to bypass small crack growth analysis and make direct use of a long crack growth law for fatigue life prediction. The equivalent initial flaw size, a_0 is calculated from material properties (ΔK_{th}, the threshold stress intensity factor and σ_f, the fatigue limit) and geometric properties (Y) as explained in Liu and Mahadevan (2008).

$$a_{0} = (1/\pi)(\Delta K_{th}\,/\,Y\sigma_{f})^{2} \qquad (5)$$

By integrating the expression in Eq. (1), the number of cycles (N) to reach a particular crack size a_N can be calculated as shown in Eq. (6).

$$N = \int dN = \int 1/(\varphi^{r} C\,(\Delta K)^{n}(1-\Delta K_{th}/\Delta K)^{m})\,da \qquad (6)$$

For structures with complicated geometry and loading conditions, the integral in Eq. (6) is to be evaluated cycle by cycle, calculating the stress intensity factor in each cycle of the crack growth analysis. The calculation of the stress intensity factor is explained in the following subsection.

3.2 Calculation of Stress Intensity Factor

The stress intensity factor ΔK in Eq. (6) can be expressed as a closed form function of the crack size for specimens with simple geometry subjected to constant amplitude loading. However, this is not the case in many mechanical components, where ΔK depends on the loading conditions, geometry and the crack size. Further, if the loading is multi-axial (for example, simultaneous tension, torsion and bending), then the stress intensity factors corresponding to three modes need to be taken into account. This can be accomplished using an equivalent stress intensity factor. If K_I, K_{II}, K_{III} represent the mode-I, mode-II and mode-III stress intensity factors respectively, then the equivalent stress intensity factor K_{eqv} can be calculated using a characteristic plane approach proposed by Liu and Mahadevan (2008) . The use of the characteristic plane approach for crack growth prediction under multi-axial variable amplitude loading has been validated earlier with several data sets.

During each cycle of loading, the crack grows and hence, the stress intensity factor needs to be reevaluated at the new crack size for the loading in the next cycle. Hence, it becomes necessary to integrate the expression in Eq. (6) through a cycle by cycle procedure. Each cycle involves the computation of ΔK using a finite element analysis represented by Ψ.

$$\Delta K_{eqv} = \Psi (a, L, \theta) \qquad (7)$$

Repeated evaluation of the finite element analysis in Eq. (7) renders the aforementioned cycle by cycle integration extremely expensive, perhaps impossible in some cases. Hence, it is necessary to substitute the finite element evaluation by an inexpensive surrogate model. Different kinds of surrogate models (polynomial chaos, support vector regression, relevance vector regression, and Gaussian Process interpolation) have been explored and the Gaussian process modeling technique has been employed in this paper. A few runs of the finite element analysis are used to train this surrogate model and then, this model is used to predict the stress intensity factor for other crack sizes and loading cases (for which finite element analysis has not been carried out).

3.3 Construction of Gaussian Process Surrogate Model

A Gaussian process (GP) response surface approximation is constructed to capture the relationship between the input variables (a, L, θ) and the output variables (ΔK) in Eq. (5), using only a few sample points within the design space. The details of this interpolation technique are available in literature (Rasmussen, 1996; Santner, 2003; McFarland, 2007).

The basic idea of the GP model is that the response values Y (K_{eqv} in this case), are modeled as a group of multivariate normal random variables, with a defined mean and covariance function. The benefits of GP modeling is that the method requires only a small number of sample points (usually 30 or less), and is capable of capturing highly nonlinear relationships that exist between input and output variables without the need for an explicit functional form. Additionally, Gaussian process models can be used to fit virtually any functional form and provide a direct estimate of the uncertainty associated with all predictions in terms of model variance. The framework of Gaussian process modeling is shown in Figure 3.

Suppose that there are n training points, x_1, x_2, x_3 ... x_n of a d-dimensional input variable (the input variables being the crack size and loading conditions here), yielding the resultant observed random vector $Y(x_1)$, $Y(x_2)$, $Y(x_3)$ ... $Y(x_n)$. R is the m x m matrix of correlations among the training points. An exponential correlation function has

been suggested by researchers in the past (Bichon et al., 2008).

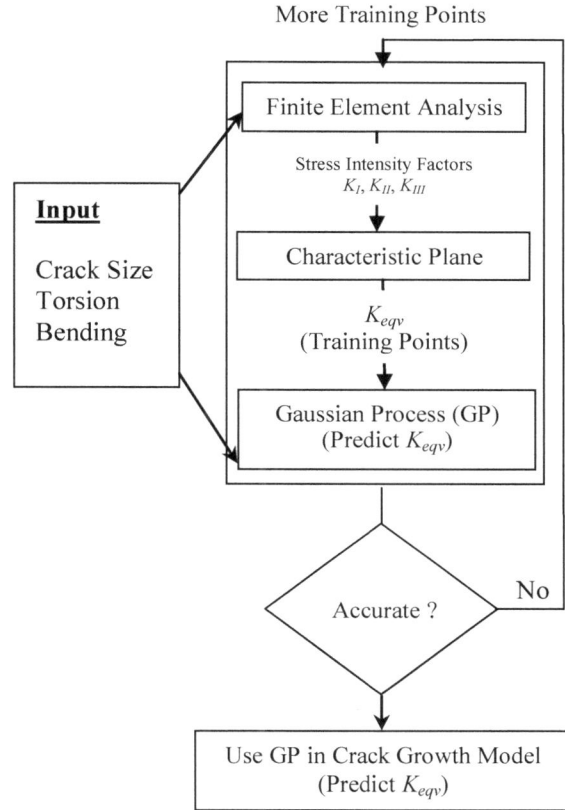

Figure 3. Construction of Surrogate Model

$$Y^* = E(Y|x^*) = f^T(x^*)\beta + r^T(x^*)R^{-1}(Y - F\beta) \qquad (8)$$

$$\sigma_{Y^*} = Var(Y|x^*) = \lambda(1 - r^T R^{-1} r) \qquad (9)$$

In Eq. (8) and Eq. (9), F is a matrix with rows of trend functions $f^T(x_i)$, r is the vector of correlations between x^* and each of the training points, β represents the coefficients of the trend function. A constant trend function has been reported to be sufficient (Sacks et al., 1989). McFarland (2007) discusses the implementation of this method in complete detail.

3.4 Crack Propagation Analysis

This section explains the method used to calculate the final crack size as a function of number of load cycles. The procedure involves the evaluation of the integral in Eq. (4). As explained in Section 3.3, this needs to be done cycle by cycle and the Gaussian process surrogate model is used to predict the equivalent stress intensity factor in each cycle. Starting with the equivalent initial flaw size a_0, the equations (Eq. (1) – Eq. (6)) described in Section 3.1 are used to calculate the final crack size A after N loading cycles. This entire procedure is summarized in Figure 4.

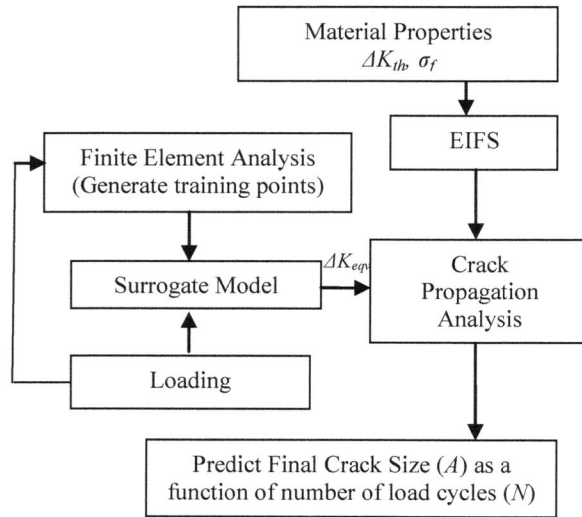

Figure 4. Crack Propagation Analysis

The framework shown in Figure 4 for crack growth prognosis is deterministic and does not account for errors and uncertainty. Uncertainty can be associated with each of the blocks in Figure 4 and accounted for in crack growth prediction. The following section investigates these sources of uncertainty and Section 5 incorporates them into the crack growth analysis methodology.

4. SOURCES OF UNCERTAINTY

This section discusses the various sources of uncertainty and errors that are part of the crack growth framework summarized in Section 3.5 and proposes methods to handle different types of uncertainty. The material properties used to calculate the equivalent initial flaw size are measured using experiments and have variability, causing variability in EIFS. Further, these experimental data may be sparse and the uncertainty in data needs to be accounted for. The crack growth law used for crack propagation is usually estimated through curve fitting of experimental data. To account for model uncertainty, a (normally distributed) error term is added to the crack growth equation and the model coefficients of the crack growth law are treated as random variables. In each cycle of loading, the stress intensity factor is calculated as a function of current crack size, loading and geometry. Repeated finite element analyses are avoided by the use of inexpensive surrogate models and the output of the surrogate model is not accurate. Further, the training points calculated using finite element analyses are prone to solution approximation and discretization errors. Further, the loading itself is considered to be random – a variable amplitude multi-axial loading case is demonstrated in this paper. These various sources of uncertainty can be classified into three different types – physical variability, data uncertainty and model uncertainty - as shown below.

I. Physical Variability

    a. Loading
    b. Equivalent initial flaw size
    c. Material Properties (Fatigue Limit, Threshold Stress Intensity Factor)

II. Data Uncertainty

    a. Material Properties (Fatigue Limit, Threshold Stress Intensity Factor)

III. Model Uncertainty/Errors

    a. Crack growth law uncertainty
    b. Uncertainty in calculation of Stress Intensity factor
    c. Discretization error in finite element analysis
    d. Uncertainty in surrogate model output

(Note: Variations in geometry and boundary conditions are sources of physical variability. These variations are not considered in this research work. However, these can be included in the proposed framework by constructing different finite element models (for different geometry and boundary conditions) and use these runs to train the Gaussian process surrogate model. Hence, these parameters are treated as inputs to the surrogate model and sampled randomly in the uncertainty quantification procedure explained later in Section 5.)

The following subsections discuss each source of uncertainty in detail and propose methods to handle them.

4.1 Physical Variability in Loading Conditions

The loading on practical structures is rarely deterministic and it is difficult to quantify the uncertainty in loading. For the purpose of illustration, variable amplitude multi-axial (bending, tension and torsion) loading is considered in this paper. A loading history consists of a series of blocks of loads, the loading amplitude being constant in each block. In this paper, the block length is assumed to be a random variable and the maximum and minimum amplitudes in each block are also treated as random variables. A sample loading history is shown in Figure 5.

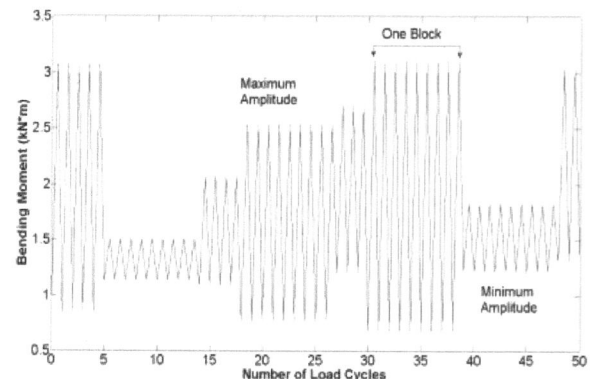

Figure 5. Sample loading history

To generate one block of loading, first a block length is selected and then a maximum amplitude value and a minimum amplitude value is selected for that block. The entire loading history is generated by repeating this process and creating several successive blocks.

4.2 Physical Variability in EIFS

The equivalent initial flaw size derived in Eq. (3) depends on ΔK_{th}, the equivalent mode-I threshold stress intensity factor, $\Delta\sigma_f$, the fatigue limit of the specimen and the geometry factor Y which in turn depends on the geometry of the structural component and the configuration of the crack. This is a deterministic quantity and can be estimated using finite element analysis. The distributions for the material properties, ΔK_{th} and $\Delta\sigma_f$ are characterized using data obtained from experimental testing. This is explained in Section 4.3. Having obtained the statistical distributions of ΔK_{th} and $\Delta\sigma_f$, the distribution of a_0, the equivalent initial flaw size, can be calculated.

4.3 Data Uncertainty in Material Properties (to characterize distributions ΔK_{th} and $\Delta\sigma_f$)

This section proposes a general methodology to characterize uncertainty in input data, from which statistical distributions need to be inferred. This method is illustrated using experimental data available in literature to characterize the distribution of threshold stress intensity factor (ΔK_{th}) and fatigue limit ($\Delta\sigma_f$). McDonald et al. (McDonald et al., 2009) proposed a method to account for data uncertainty, in which in the quantity of interest can be represented using a probability distribution, whose parameters are in turn represented by probability distributions.

Consider a random variable X whose statistics are to be determined from experimental data, given by $x = \{x_1, x_2 .. x_n\}$. For the sake of illustration, suppose that the random variable X follows a normal distribution, then the parameters (P) of this distribution, i.e. mean and variance of X can be estimated from the entire data set x. However, due to sparseness of data, these estimates of mean and variance are not accurate. Using resampling techniques such as bootstrapping method (Efron and Tibshirani, 1993), jackknifing (Efron, 1979) etc. the probability distributions ($f_P(P)$) of the parameters (P) can be calculated. Hence for each instance of a set of parameters (P), X is defined by a particular normal distribution. However, because the parameters (P) themselves are stochastic, X is defined by a family of normal distributions. For a detailed implementation of this methodology, refer McDonald et al., 2009.

Note that the aforementioned resampling techniques are useful, when considerable amount of data are available. For example, if 30 data points are needed to construct a meaningful probability distribution, then resampling techniques are not needed. Resampling techniques are very useful if there are, say 8 to 20 data points. Resampling techniques are less meaningful when there are less than 5 data points.

This paper uses resampling techniques to calculate the distribution of the parameters (P), however does not define a family of distributions. Instead, it recalculates the distribution of the random variable X, using principles of conditional probability (Haldar and Mahadevan, 2000). Thus X follows a probability distribution conditioned on the set of parameters (P). Hence the distribution of X is denoted by $f_{X|P}(x)$. However, in this case, the parameters are represented by probability distributions $f_P(P)$. Hence, the unconditional probability distribution of X ($f_X(x)$) can be calculated as shown in Eq. (10).

$$f_X(x) = \int f_{X|P}(x) f_P(P) dP \qquad (10)$$

The integral in Eq. (10) can be evaluated through quadrature techniques or advanced sampling methods such as Monte Carlo integration or Markov chain Monte Carlo Integration. Hence, the unconditional distribution of X which accounts for uncertainty in input data can be calculated. In this paper, this method has been used to characterize the uncertainty in threshold stress intensity factor (ΔK_{th}) and fatigue limit ($\Delta\sigma_f$).

4.4 Uncertainty in Crack Growth Model

There are more than 20 different crack growth laws (e.g., Paris law, Foreman's equation, Weertman's equation) proposed in literature. The mere presence of many such different models explains that none of these models can be applied universally to all fatigue crack growth problems. Each of these models has its own limitations and uncertainty. In this paper, a modified Paris law has been used for illustration, however, the methodology can be implemented using any kind of crack growth model. The uncertainty in crack growth model can be subdivided into two different types: crack growth model error and uncertainty in model coefficients. If ε_{cg} is used to denote the crack growth model error, then the crack growth law can be expressed as:

$$da/dN = \varphi^r C (\Delta K)^n (1 - \Delta K_{th}/\Delta K)^m + \varepsilon_{cg} \qquad (11)$$

An estimate of ε_{cg} can be obtained while calibrating the model parameters using statistical data fitting tools. The model coefficients in Paris law are C and n, and the uncertainty in these parameters can be represented through probability distributions. The stress intensity factor ΔK, as explained earlier is calculated using the Gaussian process surrogate model as explained in Section 3. The various sources of uncertainty in this process are addressed in Section 4.5.

4.5 Errors in Stress Intensity Factor Calculation

As explained in Section 3, a Gaussian process model is used to calculate the stress intensity factor ΔK. This is done in two stages. First, a few finite element analysis runs are required to train the GP model. Second, the GP model is used to predict the stress intensity factor as explained in Section 3.3. Each of these two steps has associated errors and uncertainty. Finite element solutions are subject to discretization errors, whereas the prediction of any low-fidelity model such as the GP model also has error. These two issues are discussed in this subsection.

4.5.1 Discretization Error in Finite Element Analysis

Theoretically, an infinitesimally small mesh size will lead to the exact solutions but this is difficult to implement in practice. Hence, finite element analyses are carried at a particular mesh size and the error in the solution, caused due to discretization needs to be quantified. Several methods are available in literature but many of them quantify some surrogate measure of error to facilitate adaptive mesh refinement. The Richardson extrapolation (RE) method has been found to come closest to quantifying the actual discretization error and this method has been extended to stochastic finite element analysis by Rebba (Richards, 1997; Rebba, 2005). It should be noted that the use of Richardson extrapolation to calculate discretization error requires the model solution to be convergent and the domain to be discretized uniformly (uniform meshing) (Rebba et al., 2004). Sometimes, in the case of coarse models, the assumption of monotone truncation error convergence is not valid. In the Richardson extrapolation method, the discretization error due to grid size, for a coarse mesh is given by Eq. (12).

$$\varepsilon_h = (f_1 - f_2) / (r^p - 1) \qquad (12)$$

In Eq. (12), f_1 and f_2 are solutions for a coarse mesh and a fine mesh respectively. If the corresponding mesh sizes were denoted by h_1 and h_2, then the grid refinement ratio, denoted by r is calculated as h_2/h_1. The order of convergence of p is calculated as:

$$p = \log ((f_3 - f_2) / (f_2 - f_1)) / \log(r) \qquad (13)$$

In Eq. (13), f_3 represents the solution for a coarse mesh of size h_3, with the same grid refinement ratio, i.e. $r = h_3/h_2$. The solutions f_1, f_2, f_3 are dependent on the inputs (loading, current crack size, aspect ratio and angle of orientation) to the finite element analysis and hence the error estimates are also functions of these input variables. For each set of inputs, a corresponding error is calculated and this error is added to the (coarse mesh) solution from finite element analysis to calculate the true solution. Hence a true solution is associated with each set of inputs and these values are used as training points for the surrogate model.

4.5.2 Uncertainty in the Surrogate Model Output

Several finite element runs for some combination of input-output variable values are used to train the Gaussian process surrogate model in this paper. Then, these surrogate models can be used to evaluate the stress intensity factor for other combinations of input variable values. GP models, as explained in Section 3.3, model the output as a sum of Gaussian variables and hence, inherently produce an output which is normally distributed. The expressions for mean and variance of the output of the GP model were given in Eq. (8) and Eq. (9) respectively. The output of the GP (ΔK_{eq}) model is a random normal variable and in each cycle, the value for ΔK_{eq} is sampled from this distribution.

(Note: The GP model is used as a surrogate for the deterministic finite element model and the variance of the GP output accounts only for the uncertainty in replacing the original model with a Gaussian process and does not account for the uncertainty in the inputs to the model. The variance of the output is only dependant on the "form" of the surrogate model. For example, a linear surrogate model will lead to constant variance at untrained locations but unknown distribution type (Seber and Wild, 1989). The advantage in using a Gaussian process surrogate model is that not only the output variance can be calculated but also the distribution type can be proved to be Gaussian (McFarland, 2007).)

The Gaussian process model output, i.e. the stress intensity factor is used in the crack growth equation to predict the crack size as a function of number of cycles as explained earlier in Section 3. The following section incorporates all these sources of uncertainty into the crack growth prediction methodology described in Section 3.

5. UNCERTAINTY IN CRACK GROWTH

Section 3 proposed a methodology that can be used for crack growth of structures with complicated geometry and subjected to multi-axial loading. This procedure was summarized using a step-by-step flowchart in Figure 4. Section 4 investigated the various sources of uncertainty in the crack growth prediction methodology and proposed methods to handle them. A brief summary of the various sources of uncertainty is given below.

I. PHYSICAL VARIABILITY

a. Variable amplitude multi-axial loading cycles are generated by considering random block lengths and random amplitudes within each block.
b. The equivalent initial flaw size (EIFS) is represented by a probability distribution that accounts for the variability in material parameters, the threshold stress intensity factor and fatigue limit.

c. The material properties (fatigue limit, threshold stress intensity factor) are represented by probability distributions, inferred from experimental data.

II. DATA UNCERTAINTY

a. The uncertainty in data used to calculate the statistics of material properties (fatigue limit, threshold stress intensity factor) is addressed by using a sampling based approach that calculates a family of probability distributions for each material parameter. Then, this family of distributions is integrated into one single probability distribution (for each property) using the principles of conditional and total probability.

III. MODEL UNCERTAINTY/ERRORS

a. The uncertainty in crack growth model is handled by adding an error term to the crack growth law and by representing the model parameters as random variables.
b. The calculation of stress intensity factor in each cycle of crack growth is facilitated using a Gaussian process surrogate model.
c. The discretization error in finite element analysis is calculated using Richardson extrapolation and added to the results of FEA before training the surrogate model.

The uncertainty (calculated as the variance) in the surrogate model output is modeled as a Gaussian variable calculation from regression results and hence, the prediction of the surrogate model, i.e. the Stress intensity factor is represented as a normal distribution.

This section presents a sampling based strategy to combine all the different sources of uncertainty and thereby quantify the overall uncertainty in crack growth prediction as a function of number of loading cycles (N). The various steps in this procedure are outlined here.

I. Generate training points for the Gaussian process surrogate model. This is done through finite element analysis and then by calculating the discretization error in each of the runs. The discretization errors are added to the solutions of finite element analysis and used to train the Gaussian process surrogate model. Hereon, the GP model can be used to calculate the stress intensity factor as a function of crack size, loading, aspect ratio and angle of orientation.
II. Generate a loading history. First, randomly select a block length and then randomly select a maximum amplitude value and a minimum amplitude value for that particular block. Repeat the process till the number of cycles (N) is reached.
III. Sample an EIFS value from the statistical distribution calculated in Section 4.1 and Section 4.2.
IV. Use the deterministic procedure for crack growth analysis to calculate the final crack size at the end of N cycles. However, in each loading cycle, the stress intensity factor calculated from the GP model is a random normal variable and hence generate a random sample of stress intensity factor in each cycle. Also, the crack growth model error (ε_{cg}) is sampled in every cycle.

In this algorithm, Step I is a deterministic step while Step II, Step III and Step IV are probabilistic. Using this algorithm, the crack size after N cycles can be calculated for a particular load history that was generated in Step II. Using Monte Carlo Sampling, Steps II, III and IV can be repeated again and again, each leading to a final crack size at the end of N cycles. This can be used to characterize the distribution of final crack size at the end of N cycles. By varying N, the distribution of final crack size can be obtained as a function of the number of cycles (N). This information can be used to calculate the reliability of the structural component as a function of number of load cycles. Suppose that the component is supposed to have failed if the crack size is greater than a critical crack size (A_c), then the probability of failure can be calculated as a function of load cycles.

Note that the proposed methodology for uncertainty quantification is applicable to any model-based technique for fatigue crack growth prognosis. Thus any appropriate combination of fatigue crack growth models can be used instead of the models used in this paper for implementing the proposed uncertainty quantification methodology.

6. NUMERICAL EXAMPLE

This section illustrates the proposed methodology to quantify the uncertainty in crack growth analysis through a numerical example.

6.1 Description of the Problem

A two radius hollow cylinder with an elliptical crack in fillet radius region is consi Coarse Full Model   This problem consists of modeling an initial semi-circular surface crack configuration and allowing the crack shape to develop over time into a semi-elliptical surface crack. This is shown in Figure 6.

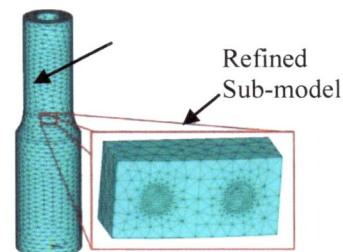

Figure 6. Surface Crack in a Cylindrical Structure

The finite element software package ANSYS (ANSYS, 2007) version 11.0 is used to build and analyze the finite element model. The crack configuration is built by extruding a projection of the semi-circular crack through the mast body at the crack location. The immediate volumes on either side of the crack face are identified and subdivided in order to allow for SIF evaluation at various locations along the crack front. The crack faces (coinciding upper and lower surfaces of the previously mentioned volumes) are then modeled as surface to surface contact elements (CONTACT174 and TARGET170 elements) in order to prevent the surface penetration of the crack's upper and lower surfaces. The augmented Lagrangian method is the algorithm used for contact simulation. Additionally, friction effect is included in the material properties of the contact element, in which a Coulomb friction model is used. This model defines an equivalent shear stress which is proportional to the contact pressure and the friction coefficient. Friction coefficients between two crack faces are difficult to measure and are generally assumed to vary between 0 and 0.5 (Liu et al., 2007). The friction coefficient, μ is assumed to be equal to 0.1.

Since the primary quantity of interest is the stress intensity factor at the crack tip, the volume along the crack front is subdivided into many smaller blocks, which allows for better mesh control and enables SIF evaluation at various locations along the crack front. The crack region is constructed within a submodel of the uncracked body. The submodel technique is based on the St. Venant's principle, which states that if an actual distribution of forces is replaced by a statically equivalent system, the distribution of stress and strain is altered only near the regions of load application. The sub-modeling technique facilitates accurate stress intensity factor solutions all along the crack front which can be used for crack growth analysis.

Table 1 and Table 2 list the material and geometrical properties of the specimen under study.

Aluminium 7075- T6	
Modulus of Elasticity	72 GPa
Poisson Ratio	0.32
Yield Stress	450 MPa
Ultimate Stress	510 MPa

Table 1 Material properties

Cylinder Properties	
Length	152.4 mm
Inside Radius	8.76 mm
Outside Radius (Narrow Sect)	14.43 mm
Outside Radius (Wide Sect)	17.78 mm

Table 2 Geometrical Properties

In reality, these parameters in Table 2 and Table 3 may be variable and might require probabilistic treatment. However,

as mentioned earlier, physical variability in the geometry of the structure, Young's modulus, Poisson ratio, boundary conditions, friction coefficient between crack faces, etc are treated to be deterministic in this paper. The following subsection discusses the numerical implementation of the uncertainty quantification procedure.

6.2 Algorithm for Uncertainty Quantification

The numerical details of the different sources of uncertainty are presented in this section. They are given step-wise in the same order as in Section 5.

I. Finite element analyses are run for 10 different crack sizes, 6 different loading cases, two angles of orientation and three different aspect ratios, amounting to 360 training points to construct the surrogate model. For each solution, three different meshes are considered and the discretization error is quantified as explained in Section 4.4.1. The discretization error is added to the finite element analysis solution at each training point and the Gaussian process model is trained to predict the stress intensity factor.

II. Multi-axial variable amplitude loading cycles are generated by considering blocks of equal amplitude within one entire loading history. The block length is assumed to be a uniform distribution (U(0,500)) and the maximum amplitude and minimum amplitude for that block are assumed to follow normal distributions (N(μ_1,σ_1) and N(μ_2,σ_2) where μ_1, σ_1, μ_2, σ_2 are uniformly distributed on the intervals [20, 28], [2, 6], [8, 16], and [2, 6] respectively, in KNm).

III. The distribution of EIFS is characterized using the data used by Liu and Mahadevan (Liu and Mahadevan, 2008). However, the current research work accounts for uncertainty in data and treats the parameters of threshold stress intensity factor and fatigue limit as random variables as well. The distribution (conditioned on its parameters) of EIFS is assumed to be lognormal (with parameters λ, ζ), with the λ following a normal distribution (mean = -7.60 and standard deviation = 0.50) and ζ following a lognormal distribution (mean = 0.22 and standard deviation = 0.10). The unconditional distribution of EIFS is calculated using the integral in Eq. (8). Samples of EIFS are drawn from this distribution.

IV. Paris law is used for crack growth propagation. The model parameter C (mean = 6.5 E-13 and standard deviation = 4E-13) is chosen to be lognormally distributed whereas m (m = 3.9) is treated as a deterministic quantity. These are identical to the distributions used by Liu and Mahadevan (Liu and Mahadevan, 2009). In each loading cycle, the values of stress intensity factor and crack growth model error (ε_{cg}) are sampled from probability distributions. While the stress intensity factor (calculated using the Gaussian process surrogate model) is a Gaussian variable (as explained in section 4.5.2), a 5% Gaussian white noise is used to represent the crack growth model error. The latter quantity is chosen to be normal (Seber and Wild, 1989) because it represents a fitting error while calculating the coefficients of modified Paris' law. Note that the true value of this noise can be estimated from actual experimental data, and a 5% value is chosen only for illustration.

Using the sampling-based framework in Section 5, the probability distribution of the final crack size is calculated as a function of the total number of cycles. A Monte Carlo simulation using 5000 runs is used to calculate the probability distribution of crack size as a function of number of load cycles. The mean, median and 90% prediction bounds of the final crack size are shown in Figure 7.

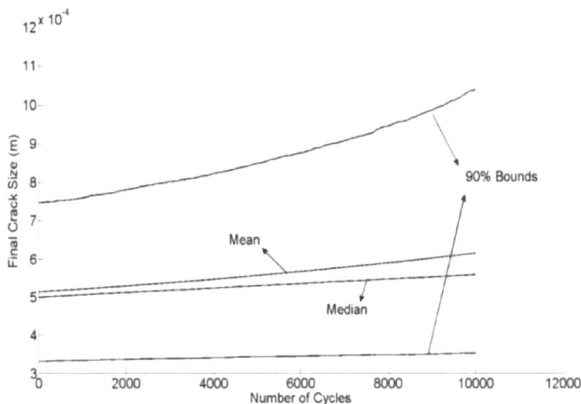

Figure 7. Mean, Median and 90% Bounds

In Figure 7, the growth of the crack is shown as a function of number of load cycles. As the number of cycles increase, there is more uncertainty and hence, the 90% prediction bounds are wider. This is due to the fact that each additional loading cycle imparts more randomness arising from variability in loading, variability in crack size at the end of previous cycle, uncertainty in the prediction of stress intensity factor, etc. To illustrate the increase in uncertainty, the standard deviation of crack size is calculated as a function of number of load cycles and plotted in Figure 8.

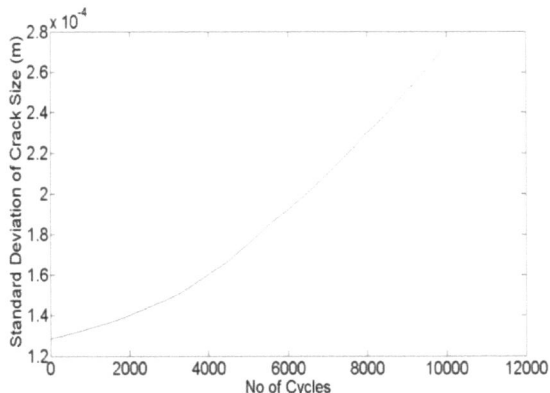

Figure 8. Standard Deviation of Final Crack Size

Figure 8 clearly shows the increase in uncertainty with number of load cycles. While the standard deviation of the initial crack size is low, it increases by about 500% at the end of 5000 load cycles. This increase is due to accumulation of different sources of uncertainty in each loading cycle, i.e. loading uncertainty, surrogate modeling errors and crack growth model errors.

Finally, the reliability of the structural component is also evaluated. A critical crack size of 2.54 mm (approximately 0.1 inch) is assumed for the purpose of illustration and the probability of failure is estimated as a function of number of load cycles and plotted in Figure 9. From Figure 9, it is seen that the probability of failure is negligible for about 3500 load cycles and it gradually increases after 4000 cycles.

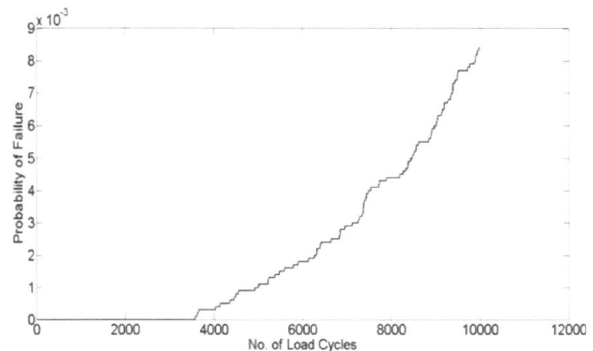

Figure 9. Probability of Failure vs. No. Load Cycles.

There are two reasons for the observed increase in increase of failure probability. Firstly, the crack is growing in size and secondly, the uncertainty in the estimated crack size also increases with each loading cycle. After 10000 cycles of loading, the probability of failure is approximately equal to 0.01.

6.3 Individual Contributions of Uncertainty

The previous subsection presented the effect of all the different sources of uncertainty in the final distribution of crack size. The current subsection calculates the marginal contributions of each source of uncertainty in the overall results of crack growth calculation. Such an analysis would identify which sources of uncertainty are critical and what the analyst must do in order to reduce the overall uncertainty in crack growth prediction.

To calculate the contribution of one particular kind of uncertainty, all other quantities are assumed to be deterministic (at their mean values) and the results of this analysis are compared with the results of Section 6.2, where all sources of uncertainty were accounted. The individual contributions of each uncertainty are tabulated in Table 3. This approach facilitates resource allocation trade-offs between model refinement and data collection for the purpose of reduction in overall uncertainty. If the contribution from the uncertainty in a particular model is high, then it is preferable to refine the model in order to reduce the uncertainty in the crack growth prediction. If the contribution from the data is high (as seen in Table 3), then it is preferable to collect more data to reduce the overall uncertainty in the crack growth prediction. Note that experimental data on material properties is used to characterize the distribution of the equivalent initial flaw size; hence, a reduction in data uncertainty would lead to reduction in EIFS uncertainty, and thus the overall uncertainty in the crack growth prediction.

Sources of Uncertainty Considered	Final Crack Size		
	Mean (mm)	Std (mm)	COV
All	0.617	0.273	0.4424
Loading	0.592	0.068	0.1152
Crack Growth Model	0.544	0.023	0.0421
Data Uncertainty	0.547	0.151	0.2767
EIFS Uncertainty	0.544	0.134	0.2463
GP Model Uncertainty	0.544	5.33E-5	9.81E-6

Table 3. Individual Contributions of Uncertainty

The above decision inferences are qualitative; future work needs to quantify the contributions of various sources of uncertainty to the overall uncertainty in the model prediction. Quantitative sensitivity analysis techniques may be pursued for this purpose.

7. SUMMARY

This paper investigated the various sources of uncertainty in a fatigue crack growth prognosis problem and illustrated the proposed methods to quantify the overall uncertainty in crack growth prediction for structures with complicated geometry and multi-axial variable amplitude loading. The concept of equivalent initial flaw size was used to replace small crack growth analysis and use a long crack growth model, specifically modified Paris law, for crack propagation. A characteristic plane approach was used to calculate an equivalent stress intensity factor in the presence of multi-axial loading conditions. Crack growth under variable amplitude loading is modeled by including retardation effects in the modified Paris law. Expensive finite element analysis was replaced by an inexpensive surrogate, i.e. the Gaussian process model, to evaluate the stress intensity factor in each cycle for use in crack growth law. Several sources of uncertainty – physical variability, data uncertainty and modeling errors - were included in the crack growth analysis procedure. Physical variability included loading conditions and material properties such as threshold stress intensity factor and fatigue limit. The uncertainty in data used to characterize these parameters was accounted for. Three different types of modeling errors – discretization errors, surrogate modeling error and crack

growth model error – were considered in this paper. A probabilistic methodology was proposed to incorporate these sources of uncertainty into the crack growth prediction methodology. A Monte Carlo based sampling approach is used to calculate the distribution of crack size as a function of number of loading cycles. By defining a suitable serviceability criterion (for example, crack size being greater than a critical value), the reliability of the structural component is calculated as a function of number of loading cycles. The methods developed for uncertainty quantification are applicable to any model-based fatigue crack growth prognosis procedure, and are not confined to the illustrative crack growth models used in this paper.

This research work also reported the individual contributions of various sources of uncertainty to the overall uncertainty in crack growth prediction. This kind of study is popularly called as global sensitivity analysis and the method presented in this paper is a heuristic approach only. Rigorous methods for sensitivity analysis have been developed by several researchers around the world and future work would involve the application of these methods to crack growth analysis problems. This study would analyze the significance of the individual contributors, assess and rank their importance, and quantify the benefits from reducing the uncertainty in the important contributors. Further, this research work considered the growth of coplanar cracks only. Non-coplanar cracks will be modeled and included in uncertainty analysis in future. Also, additional sources of variability, uncertainty, and error, such as variability in the coefficient of friction between crack surface, material properties such as Young's modulus, geometrical properties, model uncertainty in the treatment of mixed-mode cracking, and variance introduced by introduced by Monte Carlo sampling will be considered future.

ACKNOWLEDGMENT

The research reported in this paper was supported in part by the NASA ARMD/AvSP IVHM project under NRA Award NNX09AY54A (Project Monitor: Dr. K. Goebel, NASA AMES Research Center) through subcontract to Clarkson University (No. 375-32531, Principal Investigator: Dr. Y. Liu), and by the Federal Aviation Administration William J. Hughes Technical Center through RCDT Project No. DTFACT-06-R-BAAVAN1 (Project Monitors: Dr. John Bakuckas, Ms. Traci Stadtmueller). The support is gratefully acknowledged.

NOMENCLATURE

A	Crack size
N	Number of loading cycles
θ	Equivalent initial flaw size
ΔK_{th}	Threshold stress intensity factor
ΔK	Stress intensity factor in each cycle

σ_f	Fatigue limit
Y	Geometry factor
φ^r	Wheeler's retardation coefficient
C, m, n	Parameters of modified Paris' law

REFERENCES

Joint service specification guide aircraft structures, JSSG-2006. United States of America: Department of Defense. 1998.

J.P. Gallagher, A.P. Berens, and R.M. Engle Jr. (1984). USAF damage tolerant design handbook: guidelines for the analysis and design of damage tolerant aircraft structures. Final report. 1984.

A. Merati, and G. Eastaugh.(2007). Determination of fatigue related discontinuity state of 7000 series of aerospace aluminum alloys. Eng Failure Anal 2007; 14(4):673–85.

J.N. Yang.(1980) Distribution of equivalent initial flaw size. In: Proceedings of the annual reliability and maintainability symposium. San Francisco (CA): 1980.

P. White, L. Molent, and S. Barter (2005). Interpreting fatigue test results using a probabilistic fracture approach. Int J Fatigue 2005;27(7):752–67.

L. Molent, Q. Sun, and A. Green (2006) Charcterisation of equivalent initial flaw sizes in 7050 aluminium alloy. J Fatigue Fract Eng Mater Struct 2006;29:916–37.

PMGP. Moreira, PFP. de Matos, and PMST. de Castro (2005) Fatigue striation spacing and equivalent initial flaw size in Al 2024-T3 riveted specimens. Theor Appl FractMech 2005;43(1):89–99.

S.A. Fawaz (2000). Equivalent initial flaw size testing and analysis.

Y. Liu and S. Mahadevan (2008). Probabilistic fatigue life prediction using an equivalent initial flaw size distributio., International Journal of Fatigue, Volume 31, Issue 3, March 2009, Pages 476-487, ISSN 0142-1123, DOI: 10.1016/j.ijfatigue.2008.06.005.

H. Kitagawa, and S. Takahashi (1976). Applicability of fracture mechanics to vary small cracks or cracks in early stage. In: Proceedings of the 2nd international conference on mechanical behavior of materials. USA (OH): ASM International.

M.H. El Haddad, T.H. Topper, and K.N. Smith (1979). Prediction of nonpropagating cracks. Eng Fract Mech 1979;11:573–84.

F. Hemez (2005) Uncertainty Quantification and the Verification and Validation of Computational Models. Chapter in Damage Prognosis for for Aerospace, Civil, and Mechanical Systems. Edited by D. J. Inman, C.R. Farrar, V. Lopes, and V. Steffen. 2005. ISBN 0-470-86907-0. John Wiley & Sons Ltd.

D. J. Inman, C.R. Farrar, V. Lopes, and V. Steffen (2005) Damage Prognosis for Aerospace, Civil, and Mechanical Systems. 2005. ISBN 0-470-86907-0. John Wiley & Sons Ltd.

D. Chelidze and J.P. Cusumano (2004) A Dynamical Systems Approach to Systems Prognosis. J. Vib. Acoust. 126 (2). 2004. DOI. 10.1115/1.1640638.

B. Saha and K. Goebel (2008) Uncertainty Management for Diagnostics and Prognostics of Batteries using Bayesian Techniques. 2008. Proceedings of the IEEE Aerospace Conference 2008. Big Sky, Montana. Mat 1 – Mar 8, 2008.

K. Medjaher, J. Y. Moya, and N. Zerhouni (2009) Failure Prognostic by Using Dynamic Bayes Networks. In the Proceedings of the 2nd IFAC Workshop on Dependable Control of Discrete Systems. DCDS 2009. July 1 – 8, Bari, Italy.

S.W. Doebling, and F.M. Hemez (2001) Overview of Uncertainty Assessment for Structural Health Monitoring. In the Proceedings of the 3rd International Workshop on Structural Health Monitoring, September 17-19, 2001, Stanford University, Stanford, California.

F.M. Hemez, A.N. Roberson, and A.C. Rutherford (2003) Uncertainty Quantification and Model Validation for Damage Prognosis. In the Proceedings of the 4th International Workshop on Structural Health Monitoring, Stanford University, Stanford, California, September 15-17, 2003

C.R. Farrar, G. Park, F.M. Hemez, T.B. Tippetts, H. Sohn, J. Wait, D.W. Allen, and B.R. Nadler (2004). Damage Detection and Prediction for Composite Plates. J. of The Minerals, Metals and Materials Society, November 2004.

C.R. Farrar, and N.A.J. Lieven (2006) Damage Prognosis The Future of Structural Health Monitoring. Phil. Trans. R. Soc. 365, 623–632 doi:10.1098/rsta.2006.1927. Published online 12 December 2006.

R. Patrick, M.E. Orchard, B. Zhang, M.D. Koelemay, G.J. Kacprzynski, A.A. Ferri, and G.J. Vachtsevanos (2007) An integrated approach to helicopter planetary gear fault diagnosis and failure prognosis. Autotestcon, 2007 IEEE , vol., no., pp.547-552, 17-20 Sept. 2007.

S. Gupta, and A. Ray (2007) Real-time fatigue life estimation in mechanical structures. Meas. Sci. Technol. 18 (2007) 1947–1957. doi:10.1088/0957-0233/18/7/022.

S.G. Pierce, K. Worden, and A. Bezazi (2008) Uncertainty analysis of a neural network used for fatigue lifetime prediction. Mechanical Systems and Signal Processing, Volume 22, Issue 6, Special Issue: Mechatronics, August 2008, Pages 1395-1411, ISSN 0888-3270, DOI: 10.1016/j.ymssp.2007.12.004.

J.M. Papazian, E.L. Anagnostou, S.J. Engel, D. Hoitsma, J. Madsen, R.P. Silberstein, G. Welsh, and J.B. Whiteside (2009) A structural integrity prognosis system. Engineering Fracture Mechanics, Volume 76, Issue 5, Material Damage Prognosis and Life-Cycle Engineering, March 2009, Pages 620-632, ISSN 0013-7944, DOI: 10.1016/j.engfracmech.2008.09.007.

Y. Liu and S. Mahadevan (2005) Multiaxial high-cycle fatigue criterion and life prediction for metals. International Journal of Fatigue, 2005. 27(7): p. 790-80

C. Rasmussen (1996) Evaluation of Gaussian processes and other methods for non-linear regression. PhD thesis, University of Toronto, 1996

T.J. Santner, B.J. Williams, and W.I. Noltz. (2003). The Design and Analysis of Computer Experiments. Springer-Verlag, New York, 2003.

Bichon, B., Eldred, M., Swiler, L., Mahadevan, S., and McFarland, J. (2008). Efficient Global Reliability Analysis for Nonlinear Implicit Performance Functions. AIAA Journal 2008. 0001-1452 vol.46 No.10 (2459-2468). doi: 10.2514/1.34321.

J. Sacks, S. B. Schiller, and W. Welch (1989) "Design for Computer Experiments," Technometrics, Vol. 31, No. 1, 1989, pp. 41–47. doi:10.2307/1270363.

J. McFarland (2008) Uncertainty analysis for computer simulations through validation and calibration. Ph D. Dissertation, Vanderbilt University, 2008

S.A. Richards (1997) Completed Richardson extrapolation in space and time. Comm Numer Methods Eng 1997;13:558–73.

R. Rebba (2002) Computational model validation under uncertainty.Master's thesis. Nashville, TN: Vanderbilt University.

S.A. Barter, P.K. Sharp, G. Holden, and G. Clark (2002) Initiation and early growth of fatigue cracks in an aerospace aluminum alloy, Fatigue Fract. Eng. Mater. 25 (2002) (2), pp. 111–125.

A. Makeev, Y. Nikishkov, and E. Armanios (2007) A concept for quantifying equivalent initial flaw size distribution in fracture mechanics based life prediction models, Int J Fatigue (2006), Vol. 29, No. 1, Jan. 2007.

R. Cross, A. Makeev, and E. Armanios (2007) Simultaneous uncertainty quantification of fracture mechanics based life prediction model parameters, Int J Fatigue, Vol. 29, No. 8, Aug. 2007.

R. Rebba, S. Mahadevan, and S. Huang (2006) Validation and error estimation of computational models, Reliability Engineering & System Safety, Volume 91, Issues 10-11, The Fourth International Conference on Sensitivity Analysis of Model Output (SAMO 2004) - SAMO 2004, October-November 2006, Pages 1390-1397, ISSN0951-8320, DOI 10.1016/j.ress.2005.11.035

B. Efron, and R.J. Tibshirani (1993) An Introduction to the Bootstrap. Monographs on Applied Statistics and Probability 57. Chapam and Hall/CRC. 1993.

B. Efron (1979) Bootstrap Methods: Another look at the Jackknife. The annals of statistics. Vol. 7. No. 1. pp 1-26. 1979.

M. McDonald, K. Zaman, and S. Mahadevan (2009) Representation and First-Order Approximations for Propagation of Aleatory and Distribution Parameter Uncertainty. In the Proceedings of 50th AIAA/ASME/ASCE/AHS/ASC Structures, Structural Dynamics, and Materials Conference, 4 - 7 May 2009, Palm Springs, California.

A. Haldar, and S. Mahadevan (2000) Probability, Reliability and Statistical Methods in Engineering Design, Wiley, New York, 2000.

ANSYS (2007) ANSYS theory reference, release 11.0. ANSYS Inc., 2007.

Y. Liu, L. Liu, and S. Mahadevan (2007) Analysis of subsurface crack propagation under rolling contact loading in railroad wheels using FEM. Engineering Fracture Mechanics, Vol 74, pgs 2659-2674, 2007.

G.H. Besterfield, W.K. Liu, A.M. Lawrence, and T. Belytschko (1991) Fatigue crack growth reliability by probabilistic finite elements, Computer Methods in Applied Mechanics and Engineering, Volume 86, Issue 3.

G.A.F. Seber, and C.J. Wild (1989) Nonlinear Regression. New York: John Wiley and Sons.

M. Orchard, G. Kacprzynski, K. Goebel, B. Saha, and G. Vachtsevanos (2008) Advances in Uncertainty Representation and Management for Particle Filtering Applied to Prognostics. In the Proceedings of the 1st Prognostics and Health Management (PHM) Conference, Denver, CO. Oct 6-9, 2008.

P.S. Song, B.C. Sheu, and L. Chang (2001) A modified wheeler model to improve predictions of crack growth following a single overload. JSME Int J Series A 2001;44(1):117–22.

B.K.C. Yuen, and F. Taheri (2006) Proposed modifications to the Wheeler retardation model for multiple overloading fatigue life prediction. International Journal of Fatigue, Volume 28, Issue 12, December 2006, Pages 1803-1819, ISSN 0142-1123, DOI: 10.1016/j.ijfatigue.2005.12.007.

B.C. Sheu, P.S. Song, and S. Hwang (1995) Shaping exponent in wheeler model under a single overload. Eng Fract Mech 1995;51(1): 135–43.

J. Schijve (1976) Observations on the Predictions of Fatigue Crack Growth Prediction under Variable Amplitude loading. ASTM STP 595, 1976, pp. 3-23.

A.H. Noroozi, G. Glinka, S. Lambert (2008) Prediction of fatigue crack growth under constant amplitude loading and a single overload based on elasto-plastic crack tip stresses and strains, Engineering Fracture Mechanics, Volume 75, Issue 2, January 2008, Pages 188-206.

A Systematic Methodology for Gearbox Health Assessment and Fault Classification

Hassan Al-Atat[1], David Siegel[1], and Jay Lee[1]

[1] *NSF I/UCRC for Intelligent Maintenance Systems (IMS), University of Cincinnati, Cincinnati, Ohio, 45220*

atathf@mail.uc.edu
siegeldn@mail.uc.edu
jay.lee@uc.edu

ABSTRACT

A systematic methodology for gearbox health assessment and fault classification is developed and evaluated for 560 data sets of gearbox vibration data provided by the Prognostics and Health Management Society for the 2009 data challenge competition. A comprehensive set of signal processing and feature extraction methods are used to extract over 200 features, including features extracted from the raw time signal, time synchronous signal, wavelet decomposition signal, frequency domain spectrum, envelope spectrum, among others. A regime segmentation approach using the tachometer signal, a spectrum similarity metric, and gear mesh frequency peak information are used to segment the data by gear type, input shaft speed, and braking torque load. A health assessment method that finds the minimum feature vector sum in each regime is used to classify and find the 80 baseline healthy data sets. A fault diagnosis method based on a distance calculation from normal along with specific features correlated to different fault signatures is used to diagnosis specific faults. The fault diagnosis method is evaluated for the diagnosis of a gear tooth breakage; input shaft imbalance, bent shaft, bearing inner race defect, and bad key, and the method could be further extended for other faults as long as a set of features can be correlated with a known fault signature. Future work looks to further refine the distance calculation algorithm for fault diagnosis, as well as further evaluate other signal processing method such as the empirical mode decomposition to see if an improved set of features can be used to improve the fault diagnosis accuracy.

1. INTRODUCTION

Diagnosis and health assessment of rotary machinery using vibration signals from the machine has been a domain of interest for many years. Prior to a total failure, degradation and incipient level of damage in components of the machinery can demonstrate behavioral features hidden within the vibration signals. In practice, a machine with no faults will have a vibration signature, "normal" signature, based on its system dynamics and forces acting on the system. Different mechanical faults in different components will display different vibration signatures that can be differentiated from the "normal" signature, with the utilization of the proper signal processing techniques. One of the main challenges in diagnosis and health assessment of rotary machinery is the potentiality of the machine or equipment to operate in a multitude of regimes, and thus their vibration behavior will be different in each regime. Another, challenge is that the "normal" vibration signature from every operating regime might not be available for prior training. This paper proposes a multi-regime health assessment and fault-diagnosis systematic methodology for gearbox systems, which utilizes different techniques for signal processing, regime segmentation, baseline "normal" signature detection and fault-diagnosis. This methodology of developing a fault classifier without baseline data and for a system that operates under multiple regimes and loading conditions, although applied to a gearbox for this application; could be extended to other applications with proper adjustment of the selected features and regime identification method.

The 2009 Prognostics and Health Management Data Challenge (PHM 2009 Data Challenge), focused on developing and applying techniques in the area of fault classification; data collected from a generic gearbox was used to facilitate the data sets for the challenge. A schematic of the gearbox used for data collection is shown in Figure 1, note that the measured signals consisted of two accelerometer signals along with a tachometer signal. The gearbox contains several mechanical elements, including three shafts, 4 gears, and 6 bearings; the overall objective of the data challenge was to specify the condition of each of the mechanical components and specify the particular fault if it was not in the healthy state. For example, a particular bearing could be in the healthy state, or have an inner race, outer race, or ball defect. For each data set, a 45 line diagnostic output was to be specified that detailed the condition of each mechanical component based on the available vibration and speed signals.

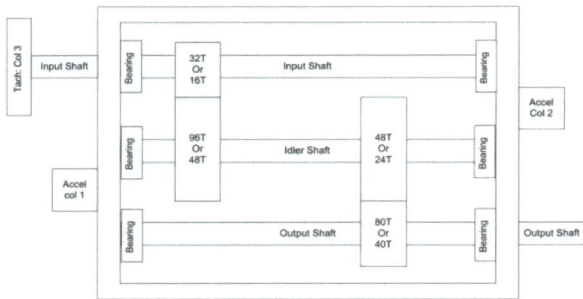

Figure 1: Schematic of Gearbox Used in PHM 2009 Challenge Data (PHM 2009 Data Challenge)

The data provided consisted of 560 data sets, in which the gearbox was tested under 5 different speeds, 2 different loads, and two different gear types. The 560 data sets consisted of data in which the gearbox was tested under different operating conditions as well as under different conditions of the mechanical components. No training data set was provided for the data challenge and fault detection was to be determined by analyzing the available vibration and speed signals and basing the diagnosis on the signature of the signal compared with the known fault signatures in the literature.

A picture of the experimental gearbox tested and used for data collection is shown in Figure 2. In this particular instance, the gearbox is shown with helical gears; however a set of spur gears was also tested and the spur gears consisted of twice the number of teeth of the helical gears.

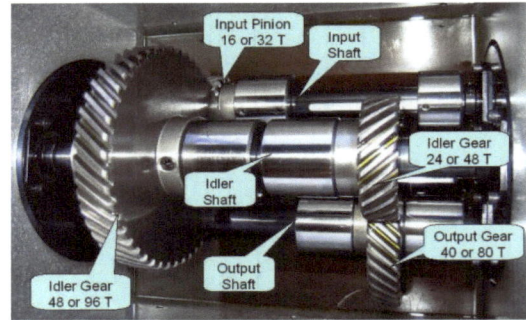

Figure 2: Inside View of Gearbox Used for Data Collection and Testing (PHM 2009 Data Challenge)

2. OVERALL METHODOLOGY TO GEARBOX FAULT DIAGNOSIS

The overall approach developed for gearbox fault diagnosis consists of several key steps in which the final output is specific diagnostic information for each mechanical component in the gearbox. For this particular application and data set, the inputs consisted of two vibration signals and a tachometer signal; the overall methodology shown in Figure 3 would require similar inputs since the fault diagnosis method is based on the vibration signals.

The initial key step requires the use of several signal processing and feature extraction methods to extract relevant information from the input vibration signals. By transforming the signal into frequency or time-frequency domains, relevant information that is correlated with particular gearbox faults can be extracted. This necessitates the use of several signal processing and feature extraction methods, since depending on the nature of the fault would indicate which signal processing method to use.

Regime segmentation allows for a fair comparison between the extracted feature sets; since the influence induced by operating the gearbox at different speeds or loads is reduced. By assessing the health and diagnosing the condition of each gearbox mechanical component for each regime; the influence of operating conditions is held constant and a change in particular features is only due to degradation of a particular component.

Health assessment consisted of using a specific set of features that are well correlated with overall gearbox health and using this feature set to assess the overall gearbox condition. For this particular application, a health assessment algorithm is used to find the data set with the minimum feature vector sum in each regime; and this is used to determine the baseline data sets.

Fault classification is the final step in the gearbox fault diagnosis methodology; the fault classification is triggered after health assessment, since only if the gearbox overall health has degraded does it necessitate further diagnosis to determine the particular problem. For each particular fault, features correlated with this failure signature are used to calculate the distance from each data set to the baseline data set in each regime, and this is used to calculate a probability of each fault based on the distance value from normal. The final diagnosis is dependent on inputs from the feature extraction step, regime segmentation, and health assessment calculation; this places much importance on the prior steps before the final diagnosis.

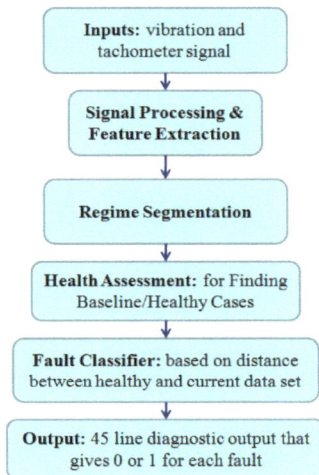

Figure 3: Flow Diagram for Gearbox Fault Diagnosis

3. SIGNAL PROCESSING AND FEATURE EXTRACTION

The health assessment and fault classification algorithms require the appropriate features as inputs in order to make the right assessment of the condition of the gearbox components as well as the level of damage. As described in the review by Samuel et al. (2005), this places much importance on extracting and selecting the most suitable set of features that are correlated to gearbox fault signatures as well as fault severity.

The signal processing and feature extraction methods used to extract a multitude of condition indicators from the vibration signals are presented; the health assessment and fault diagnosis section provide the details of the particular subset of features and algorithm used for assessing the gearbox health and providing the fault diagnosis information.

3.1 Time Domain Feature Extraction

Features from the raw time signal can be used to provide an overall understanding of the vibration level exhibited by the monitored gearbox as well as the distribution of the vibration data. The time domain feature values can be compared to a known baseline

and this provides some level of assessment of gearbox condition but limited ability to diagnosis the particular fault. The root mean square value, defined for a sampled signal is given in Eq. (1) and provides an overall indicator of the vibration energy.

$$s_{RMS} = \sqrt{\frac{1}{N} \sum_{i=1}^{N} s_i^2} \qquad (1)$$

As mentioned by Decker et al. (2003), potential time domain features include the peak to peak vibration level, crest factor, and statistical measures such as kurtosis, among others; however for the development of this particular method, only the RMS value provided a useful feature from the raw vibration time signal.

The RMS value can only provide insight that a particular fault is occurring but insight on the exact gearbox component that has damage or what failure mode is occurring cannot be inferred only using this indicator. The RMS feature can be used in an overall health assessment algorithm along with other potential features; however it has limited used for providing specific diagnosis for mechanical systems comprised of several components such as a gearbox.

An example of the level of insight the RMS feature can provide is shown in Figure 4, in which the RMS value is 2.6 times higher for the output vibration signal for a gearbox with a gear in the idler shaft having a broken tooth compared to a healthy gearbox operating under the same load and speed settings.

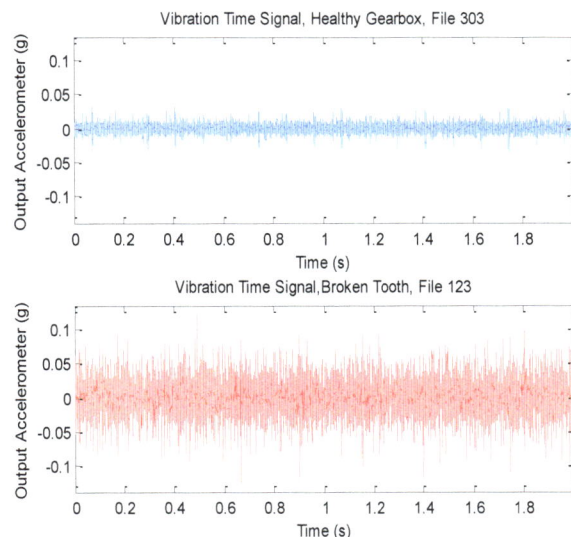

Figure 4: Vibration Time Signal for Healthy Gearbox and Gearbox with a Broken Gear Tooth

3.2 Time Synchronous Average Time Signal

By processing the tachometer signal, the vibration signal can be segmented into blocks for the duration of one revolution of the input, output or idler shaft, and averaging the signal for each block of data can highlight certain phenomena that are synchronous with the shaft rotation. Keller et al. (2003) mentioned several signal processing techniques for helicopter gearbox vibration analysis including further processing of the time synchronous signal by taking the Fast Fourier Transform and extracting peak information from particular orders of interest. However, there is still relevant information that can be extracted from the synchronous time signal including gear tooth problems related to impacts.

Impacts that occur repetitively for each shaft rotating could be an indication of a particular fault; for example a broken tooth would generate an impulsive impact once per revolution of its respective shaft and this can be more easily detected by analyzing the time synchronous average signal. As mentioned by Choy et al. (2004), the frequency spectrum obtained by performing the FFT on the time synchronous signal might not provide insight into the impact caused by this particular fault and the time or time-frequency domain is more appropriate way to analyze this particular signal. This is due to a pure impulse in the time domain containing broadband energy in the frequency domain. An example time synchronous signal is shown in Figure 5 for a data set that contains a broken gear tooth on an idler shaft gear, this particular fault has a clear signature due to the a periodic impact occurring.

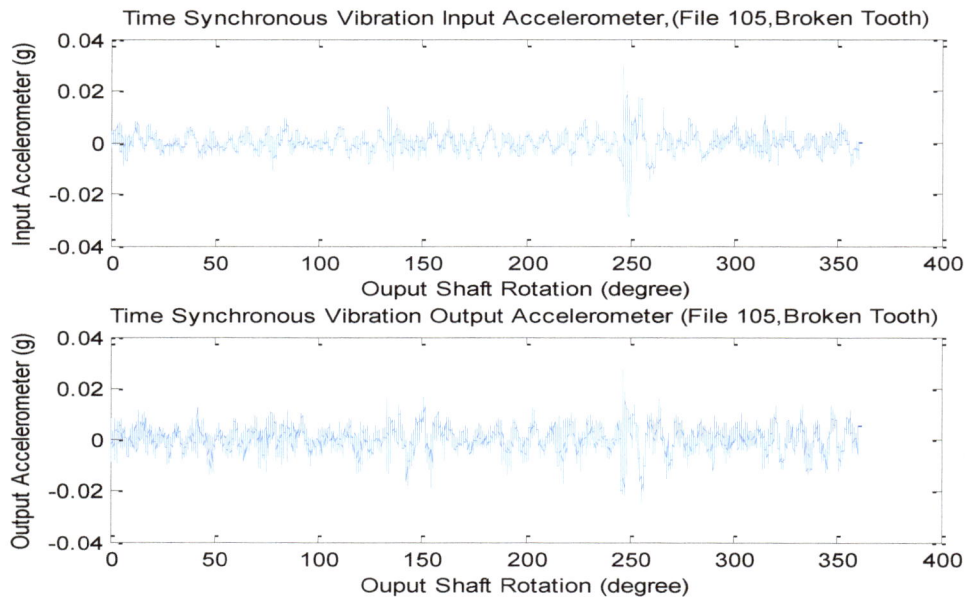

Figure 5: Time Synchronous Vibration Time Signal for Gearbox with a Gear that has a Broken Tooth

Both the peak to peak vibration and the energy operator feature can be used to process the synchronous time signal in an automated feature extraction routine to quantify this impact. The energy operator (EO) indicator used by Ma (1995), is a normalized kurtosis value defined by Eq. (2), where N is the number of data points in a sampled signal s.

$$EO = \frac{N^2 \sum_{i=1}^{N} (\Delta x_i - \Delta \overline{x})^4}{\left(\sum_{i=1}^{N} (\Delta x_i - \Delta \overline{x})^2 \right)^2} \quad (2)$$

where: $\Delta x_i = s_{i+1}^2 - s_i^2$

For the particular example shown in Figure 5, this particular gearbox with a broken gear tooth had an energy operator value that is 3 times higher than the baseline data set case for the output accelerometer signal. The energy operator in the time synchronous average signal is able to characterize this type of impact due to a gear tooth breakage. The impact could also be associated with other gearbox faults related to bearings, bent shaft, or bad key; however quantifying the impact is important and more than one feature can be used to isolate a particular fault.

3.3 Synchronous Average Vibration Spectrum
As discussed by Grabill et al. (2001), the Fourier Transform of the time synchronous averaged signal can reveal periodic occurrences that are related to a particular shaft of interest; information related to shaft

problems such as imbalance, as well as gear problems related to sidebands can be analyzed using the frequency domain spectrum. Figure 6 shows the frequency domain spectrum for a healthy gearbox and a gearbox with input shaft imbalance.

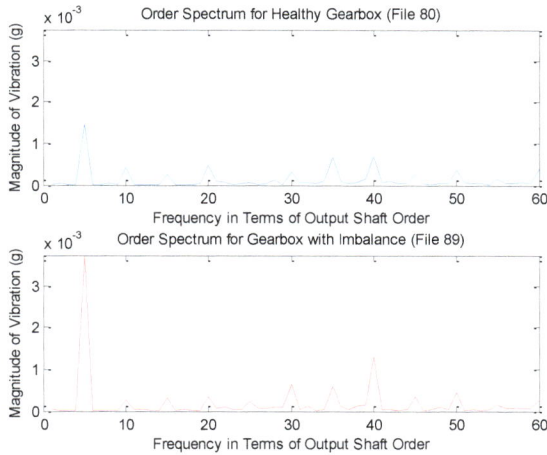

Figure 6: Frequency Spectrum for Healthy Gearbox and Gearbox with Imbalance Input Shaft

The frequency spectrum plot shows a much higher peak in the spectrum at 5X, which is the speed of the input shaft. For this particular example, the vibration from the input accelerometer at 5X is 2.5 times greater for the gearbox with imbalance in the input shaft compared to the healthy gearbox.

3.4 Continuous Wavelet Transform

For visual understanding of the impulse in the time synchronous signal, time-frequency method such as the continuous wavelet transform can reveal the broken gear tooth impact as described by Zheng et al. (2002). In this particular example, the continuous wavelet transform with a mother wavelet of Daubechies order 8 is used to process the time synchronous average signal and the impulse can be clearly seen in Figure 7. For faults that cause impacts or are transient in nature, the use of time-frequency signal processing methods such as the continuous wavelet transform can be used to provide a visual understanding of the particular fault that is occurring.

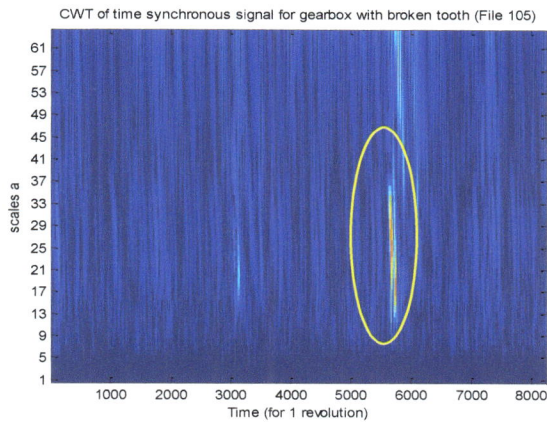

Figure 7: Continuous Wavelet Transform for Gearbox containing a gear with a broken tooth

3.5 Discrete Wavelet Transform

The vibration exhibited by a gearbox with a multitude of frequency components requires the use of advanced signal processing techniques to decompose the signal to isolate particular fault signatures more easily. The use of wavelet decomposition described by Peng et al (2003) is well suited for this particular task, in that the high frequency aspects of the signals denoted as the details can be isolated from the low frequency components. The gearbox vibration signal for a bent shaft shown in Figure 8 is an example in which the use of the wavelet decomposition technique provided a more robust feature set for fault diagnosis.

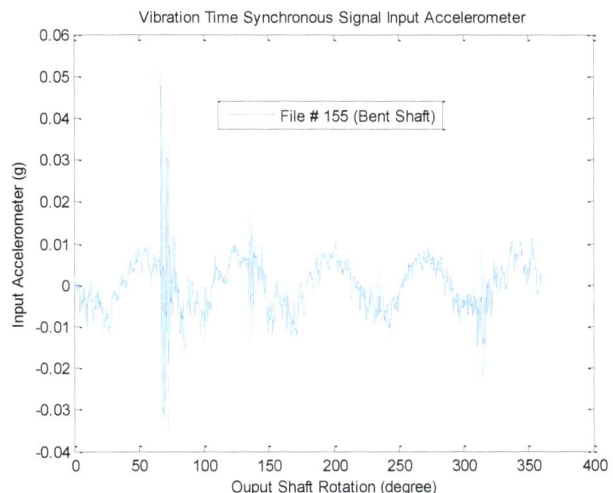

Figure 8: Vibration Time Synchronous Average Time Signal for Gearbox with Bent Input Shaft

The time synchronous vibration signal for a bent shaft condition shows an impulse impact along with a lower frequency harmonic component. The impact can be quantified by features such as kurtosis from the time synchronous signal; however the harmonic component is of particular interest since it is occurring at a frequency of 5 times the output shaft rotating speed which is the input shaft speed.

A wavelet decomposition of level 5 with a mother wavelet of Daubechies order 8 is used to further decompose the signal; the result in Figure 9 shows the approximation level 5 signal and the original signal after removing the approximation signal. The decomposition technique can be used to further analyze the harmonic component or the impact.

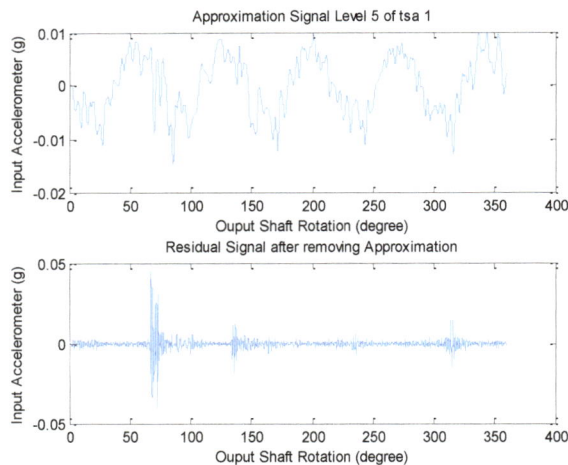

Figure 9: Wavelet Decomposition Signal for Bent Shaft Case

Further processing of the harmonic signal by taking the Fast Fourier Transform shown in Figure 10 can also be used to extract additional information.

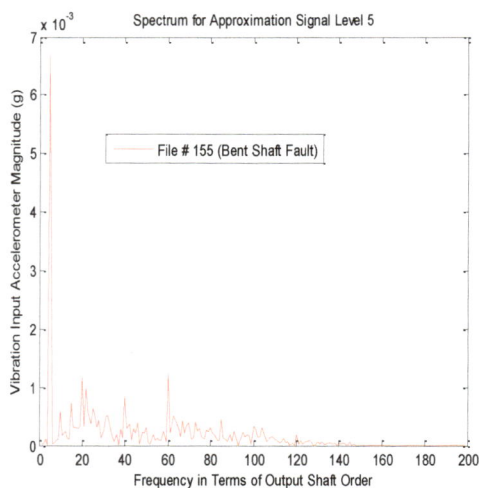

Figure 10: Frequency Spectrum of Harmonic Component after using wavelet decomposition

The frequency spectrum prior to decomposition would be difficult to interpret since the impulse would spread broadband energy; the frequency spectrum of the approximation signal provides much more relevant information related to the harmonic component associated with the input shaft speed. The peak at 5X divided by the next largest peak in the approximation signal spectrum was one of the features used to characterize the input bent shaft fault.

3.6 Spectral Kurtosis

Applying the kurtosis statistical measure for the raw vibration time signal is not necessarily suitable for detecting incipient damage and the use of a more localized way of capturing the transient nature of impulses generated from gears or bearings with early stages of damage are needed. The use of the spectral kurtosis by Antoni et al. (2006) has shown to be suitable solution for characterizing the transient impulsive type faults that occur for mechanical components; the spectral kurtosis technique has shown to be effectively used as a machine surveillance indicator as well as a way to select an optimum band pass filter for mechanical fault detection.

The overall procedure for computing the spectral kurtosis described by Antoni (2006) consists of taking the Short Time Fourier Transform (STFT) for a given block-size and computing the kurtosis statistical calculation across each spectral line; this provides a kurtosis value as a function of the frequency. The use of spectral kurtosis for the purpose of the gearbox health assessment and diagnostic method was to use the features from the spectral kurtosis calculation to characterize the overall health status of the gearbox.

A plot of the spectral kurtosis value as a function of frequency is shown in Figure 11 for a healthy gearbox and a gearbox with a broken tooth on an idler shaft gear. There is a noticeable difference in the kurtosis value at higher frequency and in particular for the output accelerometer in a frequency range of 10-20 KHz, the kurtosis value is much larger for the damaged gearbox.

For both the input and output vibration signal, three frequency bands (below 10 KHz, from 10 KHz-20 KHz and above 20 KHz) were used to calculate the sum of the kurtosis value in each frequency band and these 6 features were potential features that could be selected to determine the overall gearbox health.

In this particular example shown in Figure 11, the spectral kurtosis sum feature from 10 KHz-20 KHz for the output accelerometer was 9.5 times greater than the same feature for the healthy gearbox; this reaffirms the utility of the spectral kurtosis feature extraction method for gearbox healthy assessment. The spectral kurtosis band features provide an additional indicator for determining the overall health state of the gearbox but does not provide detail information on the particular fault that is occurring; further diagnosis requires additional features to isolate a particular gearbox fault.

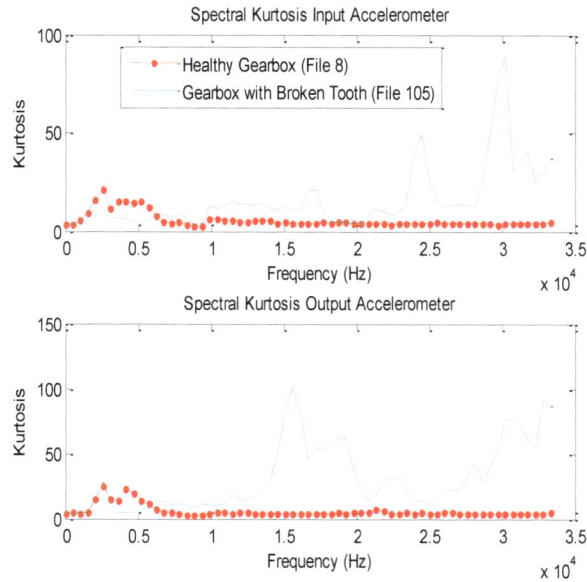

Figure 11: Spectral Kurtosis Plot for Healthy Gearbox and a Gearbox with Broken Gear Tooth

3.7 Envelope Spectrum for Bearing Fault Frequency Peak Information

The high frequency envelope frequency extraction method is a well established method for providing fault information for a particular bearing of interest. The overall signal processing procedure consists of band-pass filtering around an excited natural frequency, using the Hilbert Transform to calculate the envelope, and taking the Fourier Transform of the analytical signal. As described by Tse et al. (2001), the impacts caused by a bearing defect excite a few high frequency modes of the system and the bearing fault frequencies are amplitude modulated; by performing the band pass filtering and demodulating the signal, this characteristic fault information at the peaks can be extracted. For this particular application, a Chebyshev band pass filter was centered at 8950Hz with an upper frequency limit set at 9250Hz and a lower frequency limit at 8650Hz. For each particular bearing fault such as an inner race or outer race defect, there is a particular peak in the frequency domain that is representative of damage for this particular bearing failure mode. As mentioned in Li et al. (2000), from the specified bearing geometry, there is a set of equations that relate the bearing fault frequencies with the number of rolling elements, pitch diameter, contact angle, and diameter of the rolling elements. Figure 12 shows the envelope spectrum for the output accelerometer for both a healthy gearbox and a gearbox with an input shaft (output side) inner race bearing defect. The peak at 245Hz, which corresponds to the ball pass frequency inner race (BPFI), is clearly much higher for the gearbox with this particular inner race problem. In a similar manner the peaks at the other bearing fault frequencies can be extracted from the envelope spectrum and this subset of features can be used as inputs to classify the condition of the gearbox bearings.

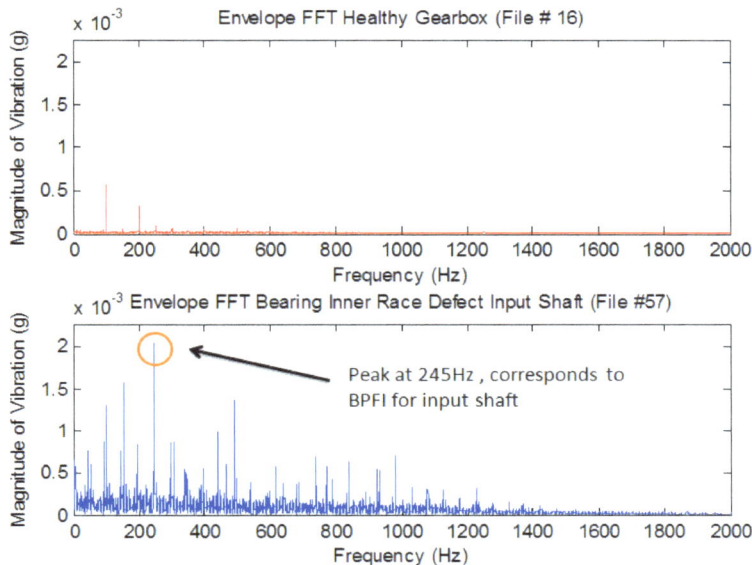

Figure 12: Envelope Spectrum for Healthy Gearbox and Gearbox with Inner Race Bearing Defect

3.8 Features Specifically for Gear Fault Diagnosis

Particular features have been specifically developed for monitoring the health of gearbox gears, and these specific gear condition indicators were extracted and considered for use in the health assessment and fault diagnosis algorithm. A more detailed description of the FM4, FMO, NB4, sideband level and index indicators are provided by Vecer et al. (2005), and a quick review of the potential use of each gear condition indicators is presented since the use of these indicators were incorporated into the gearbox health assessment and classification algorithm. Higher sidebands around the gear-mesh frequency can indicate wear or manufacturing error such as eccentricity, the sideband index is an average value of the sidebands for a particular gear mesh frequency and this feature was taken for each gear. The sideband level is a similar feature and is a ratio between the sum of the sidebands around a particular gear mesh frequency divided by the standard deviation in the time average synchronous signal.

The FM4 is taken from the residual signal, in which the gear-mesh harmonics and shaft harmonics are removed and the kurtosis is calculated for the residual signal; if one tooth is detective or damaged this feature should be greater than normal. The NB4 feature is also used to characterize gear damage and consist of band pass filtering around a particular gear mesh frequency and taking the envelope of the signal using the Hilbert Transform and calculating the kurtosis for the analytical signal. The FMO feature, also known as the zero order figure of merit, is calculated by taking the ratio between the peak to peak vibration levels for the time synchronous average signal divided by the sum of the gear mesh harmonics. These particular features were potential features used in the health assessment and fault classification method.

4. REGIME SEGMENTATION

In order to assess the gearbox condition and diagnosis particular faults, it is necessary to minimize the effect of operating variables. This allows for a fair comparison, because the operating effects would influence the features extracted from the vibration signal and higher feature values might only be due to loading or speed effects and not due to degradation in a particular gear, bearing or shaft component. The overall regime segmentation method is shown in Figure 13 and is used to segment each data set by gear type, load, and speed. The result of the regime segmentation procedure is 20 clusters, where each data set is segmented by load, speed, and gear type.

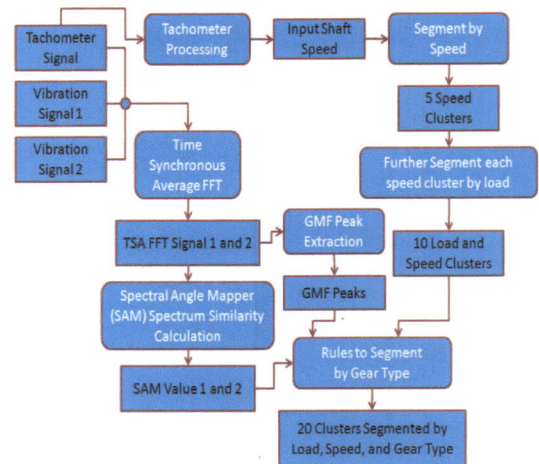

Figure 13: Regime Segmentation Flow Chart

4.1 Segment by Speed and Load

The data sets provided from the gearbox test-rigs were collected during different operating regime settings, including different input shaft speeds as well as two levels of applied braking torque load on the output shaft. By processing the tachometer square wave signal, the input shaft speed can be determined; for this particular experimental testing it is clearly observed that the input shaft speed was tested at 5 different speeds. A light or heavy braking load was applied to the output shaft, depending on the loading case. Further analysis of the speed signal in each regime showed two clusters, the cluster that had the slightly lower speed was due to a greater braking torque load being applied. Figure 14 shows the input shaft speed for the 45Hz input shaft speed cluster, the higher braking torque load causes a slight reduction in input shaft speed and allows for segmenting each data set into a high or light load regime.

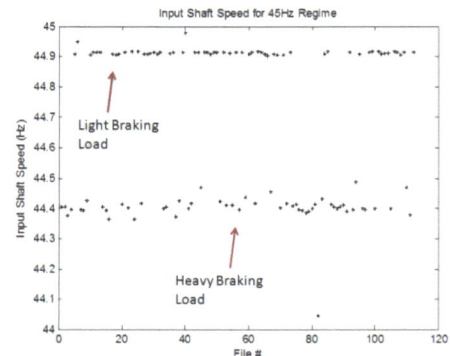

Figure 14: Example of Segmenting by Load

4.2 Similarity Spectrum Measure for Gear Type Segmentation

The gear mesh frequency is based on the shaft speed and the number of teeth on a particular gear; the helical gears used in this gearbox had a gear mesh frequency at 40 and 80 times the output shaft speed while the spur gears which had twice as many teeth had a gear mesh frequency at 80 and 160 times the output shaft speed. Using only information at the gear mesh frequency peaks at orders of 40, 80, and 160, was not enough information to segment the data sets by gear type. This is due to harmonics of the gear mesh frequency would coincide with the gear mesh frequency of the other gear type; additional information is necessary to segment the data set by gear type.

A similarity measured defined as the Spectral Angle Mapper has been used by Sheeley et al (2009) for current spectrum signals, and is incorporated into this gear type segmentation task. The time synchronous average spectrum between two data sets denoted by s_i and s_k is compared using the similarity measure shown in Eq. (3). This calculation is essentially a dot product calculation and is a value between 0 and 1, with 1 indicating a pair of spectrum that is closer in similarity.

$$SAM(s_i, s_k) = \left(\frac{dot(s_i, s_k)}{|s_i| \|s_k\|} \right) \qquad (3)$$

The use of the similarity measure for segmenting by gear type is outlined by the procedure listed below:

1. Find the data set in each of the 10 speed and load clusters that have the maximum value of the sum of the vibration spectrum peak at orders of 40 and 120 times the output shaft speed from the output accelerometer.
2. Calculate the similarity measure between this data set and all the other data sets in that regime, and take the 24 data sets that have the most similar spectrum as the helical gear type.
3. The other remaining data sets in that regime are given the label as spur gears.

By utilizing the information that corresponds to the gear mesh frequency peak and 3 times the gear mesh frequency peak for a helical gear mesh pair for the idler and output gear mesh pair, the data set with this maximum sum in each speed and load cluster is found. A data set in a particular regime that has a very similar spectrum would imply that it also contains helical gears. An example plot is shown in Figure 15 (a), in which 24 helical gears are clearly separable using this method.

This method for segmenting files by gear type was validated for a new labeled dataset published by PHM

society for the same gearbox test-rig (PHM public Datasets, 2009). The labeled dataset consisted of 14 cases, in which 8 were from spur gear and 6 from helical gear. Each labeled case, was run in 5 different speeds (30Hz, 35 Hz, 40Hz, 45 Hz, and 50Hz) under two loads (high and low) and 2 replications. Figure 15 (b) shows the segmenting by gear type for the labeled data in one regime; however the method showed 100% classification for all other regimes.

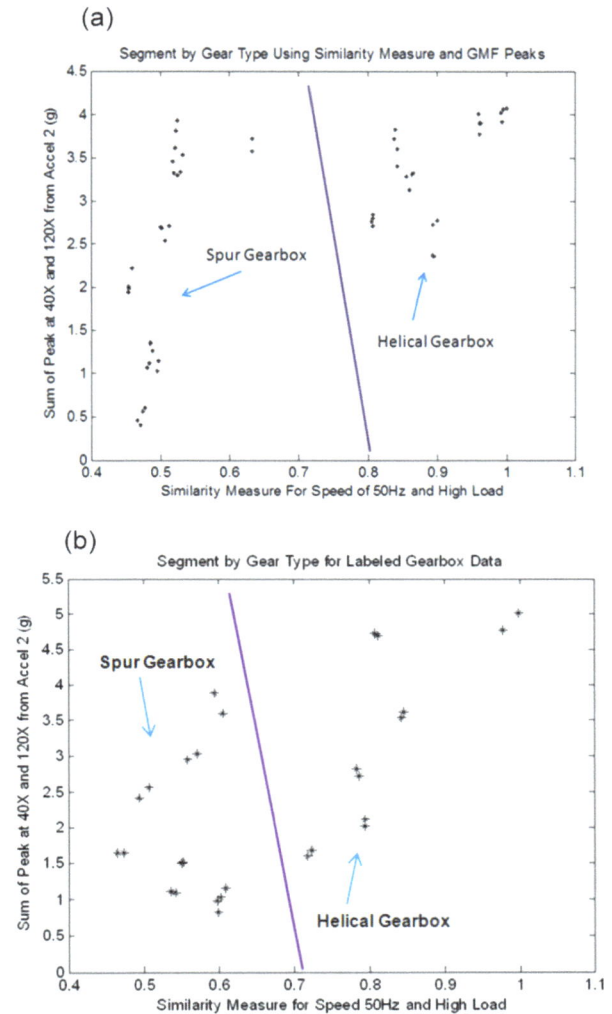

Figure 15: Example of Segmenting by Gear Type for 50Hz and high load regime a) for PHM data challenge 2009 dataset (PHM 2009 Data Challenge) b) validated result for labeled dataset (PHM Public Datasets, 2009)

5. GEARBOX HEALTH ASSESSMENT

An overall system health method or anomaly detection routine is used to provide an initial measure of diagnosis; it determines whether the system is in the normal state but not information on what fault is occurring. Although knowledge of the exact fault that is occurring is useful for reducing the maintenance cost related from logistics of ordering spare parts as well as reducing the labor time to determine the problem; only if the system health is degrading does it make sense to trigger a fault diagnosis classifier. In this particular instance, an overall system health calculation is done to determine the data sets in which the gearbox is in the baseline healthy state regarding all of its components; this baseline data set in each regime is later used by the fault diagnosis calculation.

5.1 Feature Set for Overall Gearbox Health

A list of the feature set is provided below and includes features related to overall vibration level, gear sideband information, features that are correlated to impacts from broken tooth, bearing fault frequency features, and features related to peaks due to shaft imbalance or other shaft problems.

1. RMS Value from raw time signal (input and output accelerometer).
2. Peak to Peak Level from time synchronous signal (input and output accelerometer).
3. Energy Operator from time synchronous average signal (input and output accelerometer).
4. Peak at 25X from input accelerometer from time synchronous average FFT.
5. Peak at 5X from input and output accelerometer from time synchronous average FFT.
6. Mean, max and sum of a set of features related to peaks corresponding to sidebands around gear mesh frequency.
7. Mean and sum of a set of the set of features related to sideband index and sideband level.
8. Mean of spectral kurtosis features.
9. Mean of a set of features related to bearing fault frequency peaks.
10. Max of a set of features related to peaks for shaft related problems (10X, 15X, 20X).

5.2 Health Assessment Calculation

For each of the 20 regimes segmented by gear type, load and input shaft speed, the following health assessment procedure was used to determine the overall gearbox health state as well as find the baseline data sets.

1. Normalize the feature set for health assessment so each feature has the same weight.
2. Calculate the sum squared of the feature vector for each file and store this value as the health value for this data set.
3. In each regime, find the file that has the minimum health value, this gearbox data set would be in the best health state since it has low

vibration level, and features related to gear and bearing and shaft problems are all low values.

4. After finding the gearbox data set with the minimum health value in each regime, use the similarity measure to find the other 3 data sets that are most similar to this healthy one.
5. For each operating regime, there are 4 replications, so the similarity measure is just used to ensure that the other 3 data sets that are also healthy in each regime are included.

Overall, this method was able to find all 80 data sets for healthy gearbox; the baseline data sets were later used for designing the fault classification algorithm.

Baseline Data files	Speed	Load	Gear Type
14 , 428, 431, 490	30 Hz	Light	Spur
84 , 379 , 391 , 439	30 Hz	Light	Helical
175 , 190, 369 , 446	30 Hz	Heavy	Spur
72 , 287, 497, 531	30 Hz	Heavy	Helical
101 , 165, 412, 524	35 Hz	Light	Spur
70 , 380, 420, 548	35 Hz	Light	Helical
184 , 265 , 355, 543	35 Hz	Heavy	Spur
108 , 238 , 436 , 463	35 Hz	Heavy	Helical
44 , 209 , 322 , 441	40 Hz	Light	Spur
8 , 77, 182, 252	40 Hz	Light	Helical
42 , 95 , 113, 469	40 Hz	Heavy	Spur
233 , 297 , 320 , 444	40 Hz	Heavy	Helical
181, 222, 356, 462	45 Hz	Light	Spur
303, 504, 505, 519	45 Hz	Light	Helical
116 , 212, 350, 425	45 Hz	Heavy	Spur
29 , 128 , 186 , 452	45 Hz	Heavy	Helical
4 , 172, 324 , 347	50 Hz	Light	Spur
16 , 94, 194, 460	50 Hz	Light	Helical
80, 193, 404, 555	50 Hz	Heavy	Spur
376, 481, 526 , 536	50 Hz	Heavy	Helical

Table 1: Identified baseline data files in each regime

6. GEARBOX FAULT DIAGNOSIS

6.1 Fault Diagnosis Overall Method

The fault diagnosis process is responsible for identifying the location and kind of defects in each data set. There are different techniques that can be used for this purpose; with each technique having its own merits and drawbacks. Occam's razor principle indicates that "the simplest method" to model the problem should be preferred. In this case, the simplest diagnosis method is the rule-based diagnosis where each defect is diagnosed based on the values of some features exceeding specified thresholds. The drawback of rule-based is that the specification and selection of the thresholds will be very problem specific and cannot be generalized for other problems, especially since this particular application is dealing with physical defects. Although the vibration signature could usually indicate degradation and defects, it is quite difficult to specify thresholds for specific defects that can be generalized with a high level of certainty.

Given the limitations of the rule-based diagnosis model in finding generalized thresholds or rules, a more general approach should be pursued. The approach should be capable of providing a more general method that provides the desired level of accuracy regarding fault classification.

The overall proposed approach for gearbox diagnosis is a systematic regime-and-similarity-based and is summarized in Figure 16. The diagnosis will be performed in each regime separately and will be based on calculating the probability of defect for each of the defects (33 total defects) from the distance to the regime's baseline. For each kind of defect a specific set of features are selected (based on experience and established literature for specific faults); the selected features should be capable of distinguishing the fault; the feature set will be normalized (0 to 1 scale); the distance of the feature set of each data file to the baseline in each regime is calculated; the Probability of defect P(d) is calculated for each data file for each of the 33 defects; and finally the probability value will be used to determine the existence of each defect in the data set.

The proposed approach was used to detect 5 different defects: Broken tooth in idler shaft gear output side (Gear 3); input shaft imbalance; input shaft bent; output shaft bad key; and inner race defect for bearing on input shaft output side. For each defect, different features were selected based on experience and the established literature in which certain features are known to be correlated to defects or degradation of specific mechanical components. Other faults in other components can also be detected using the same methodology; only different features should be selected based on the particular fault signature.

For the specific application of diagnosis using vibration data, most defect features are expected to increase from normal level if the defect exists. But, if the defect does not exist, the specific features are expected to be either equal or within the variation of the normal set. Since there is not enough data to establish the variations of the normal set, any feature value below the normal feature value will be considered within the normal

region of operation, and any value greater than the normal feature shall be considered outside the normal region and might be an indicator of a defect based on how far away it is from the normal region.

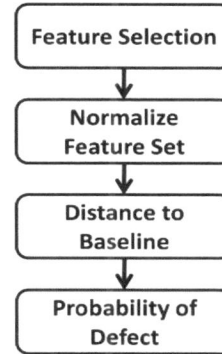

Figure 16: Overall diagnosis approach

The distance function used is a modified Euclidean distance function as shown in Eq. (4-5) as follows:

$$D(X,Y) = \left\{ \sum_{j=1}^{n} \left[\left(X_j - Y_j \right)^2 U \left(X_j - Y_j \right) \right] \right\}^{\frac{1}{2}} \quad (4)$$

$$U(X_j - Y_j) = \begin{cases} 0 & \text{if } X_j < Y_j \\ 1 & \text{if } X_j \geq Y_j \end{cases} \quad (5)$$

Where D(X,Y) is the distance between the two feature vectors X and Y each of size n, and U(X-Y) is the unit step function as defined by Eq. (5). The modified distance function will be used to find the distance of each feature vector in a regime to the feature vector of the normal set. But, some data sets might have some features lower than the normal set, which can be explained by the variations in the normal regime operation. These data sets are within the normal variation for the specific feature. The variance in a particular vibration feature can be due to several factors related to sensor noise and for this application the noise on both accelerometers and the tachometer signal would have an influence on the feature values. The probability of defect calculation, by using a distance measure that combines information from multiple features and sources of information as well as not considering features that have values less than the baseline value accounts for some of the uncertainty in the feature values due to sensor noise.

The traditional Euclidean distance measure uses the difference between the attributes squared, and thus if a feature is less than normal its distance is still positive and will be considered an increased value with respect to normal; and hence would not be differentiated from a feature that is higher than normal with the same difference. The modified Euclidean distance will consider any feature value below the normal value to be within the normal region and the distance of the specific feature will be zero. Thus the modified distance function D(X,Y) ,presented in Eq. (4), can be considered as the distance from feature vector X to the region bounded by Y rather than the distance from feature vector X to feature vector Y. Giving a weight of zero to features that are below normal or a set threshold in a distance calculation shares some similarities to the non-linear mapping method described by Bechhoefer et al. (2003), used in the vibration based health indicator calculation for the helicopter health and usage monitoring system.

The probability of defect $P_f(d)$ for a given data set f is calculated as shown in Eq. (6).

$$P_f(d) = 1 - \exp^{-D\left(X_f, N_f\right)} \tag{6}$$

Where $P_f(d)$ is the probability of the data set having a defect type d (where d=1,…33); X_f is the normalized feature vector for data set f; N_f is the normalized feature vector for the normal baseline in the same regime of data set f; and $D(X_f,N_f)$ is the modified Euclidean distance function described in Eq. (4). Since in each of the 20 regimes (shown in table 1) there are four normal data sets, then N_f is the mean of the normalized feature vectors of the four normal data sets in that regime. Note that $P_f(d)$ would be a value between zero and one, indicating the probability of each data file of having a specific defect, based on its distance from the normal baseline within its regime. The diagnosis process for each defect type will be based on selecting the appropriate feature vector that is able to isolate the defect from other possible mechanical defects.

6.2 Gear Tooth Breakage Signature
The proposed method was used to diagnose broken tooth problem in the idler shaft gear that was meshing with the output gear. The expected signature of a gearbox with a broken tooth is an overall higher level of vibration energy, impact occurring once per revolution of the broken tooth in the time signal; higher sidebands around the Gear Mesh Frequency (GMF); and natural frequency excitation due to the impact from the broken tooth. Due to the complexity of the given problem and the overlap of different defect signatures, these signatures could also be an indicator of other problems. A subset of features is needed that can

isolate this specific defect from other defects that could have similar signatures.

The selected features for this case are provided below:

1. RMS of envelope signal of the input accelerometer.
2. RMS of envelope signal of the output accelerometer.
3. Ratio between RMS of raw signal from output accelerometer to input accelerometer.
4. Ratio between sidebands: (sum of sidebands around gear 3) divided by (sum of sidebands around gear 4) from output accelerometer.

The first two features help detect the overall vibration excitation around the natural frequency; this is quantifying the natural frequency excitation due to the impact defect. Although a bearing defect will also excite high frequency modes of the system; the energy in the envelope spectrum would be at a few peaks and not spread across the entire spectrum for a bearing defect. The third feature indicates that the vibration on the output side is generally higher than the input side (pointing towards either Gear 3 or Gear 4 problems rather than Gear 1 or Gear 2). The fourth feature indicates that the sidebands of Gear 3 are higher than those of Gear 4 and points towards Gear 3 as the probable cause. The combination of these four features indicates the signature of a gear with broken tooth problem; on the output side; and most likely due to a Gear 3 problem.

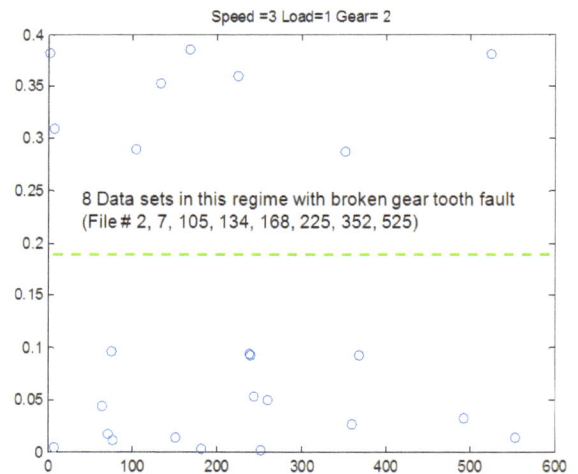

Figure 17: Probability of Gear 3 broken tooth defect for 40Hz speed, light load, and helical gear regime

After calculating the probability of defect for each data set in its regime using the four features described above, it was clear that in each regime 8 data files could be separated from the other data files. Figure 17, shows an example from a regime (speed= 40 Hz, Load=Light, Gear= Helical) where the probability of defect was calculated, and it is clear that eight of these data files can be isolated from the other files because they have a higher probability of defect for a broken tooth in Gear 3. The same method could be used for detecting a broken tooth problem in the other gears or for diagnosing other gear problems, contingent upon selecting the appropriate feature set.

6.3 Shaft Imbalance Fault Detection

The proposed method was also used to diagnose impact shaft imbalance. A mass imbalance for a particular shaft would have a higher peak corresponding to 1X for the particular imbalance shaft and this particular indicator is commonly used to diagnosis shaft imbalance problems for rotating machinery. But for this specific problem, the peak at 1xrpm also overlaps with the peaks at the characteristic bearing fault frequencies; in particular the BPFO of the bearings on the idler shaft (1.0174xrpm of input shaft) and the BPFI of the bearings on the output shaft (0.9895xrpm of input shaft). So, a high amplitude peak at 1xrpm of input shaft could be caused by either of these problems. The selected features are as follows:

1. Peak in time synchronous average spectrum from input accelerometer at input shaft speed.
2. Peak in time synchronous average spectrum from output accelerometer at input shaft speed.
3. Peaks in envelope spectrum at input shaft speed from both input and output accelerometer.

The first two features are from the time synchronous average and should filter out non-synchronous multiples of 1xrpm of input shaft. The third feature indicates amplitude modulation caused by the imbalance. This provides a feature set that provides indication of an input shaft imbalance but also indicators that are not influenced from other potential mechanical defects that have peaks in a similar frequency range.

After calculating the probability of defect for each data file in its regime using the three features described, it was clear that in each regime, 4 data sets could be separated from the other data sets. Figure 18, shows an example from a regime (speed= 50 Hz, Load=Heavy, Gear= Spur) where the probability of defect was calculated, and it is clear that four of these data files can be isolated from the other files because they have a higher probability of defect for input shaft imbalance.

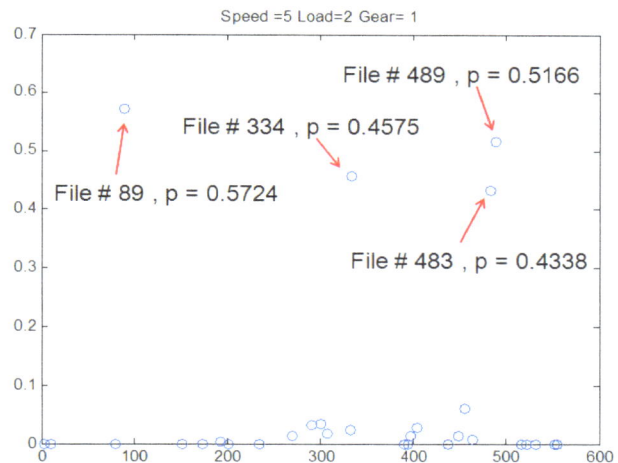

Figure 18: Probability of input shaft imbalance defect for regime of 50Hz speed, heavy load, and spur gears.

6.4 Bent Shaft Diagnosis

A bent shaft signature is dependent on the location of the bent, i.e. bent before the bearings, bent on the bearings, bent on the gears, or bent on the couplings. Each one of the aforementioned cases has a distinct signature. A common signature characteristic of all these potential bent shaft faults is the high amplitude at the 1X harmonic of the shaft speed. Specific additional characteristics would apply to the other cases, such as a bent on the coupling which would have an impact in the time domain signal which was the case for this data.

For detecting a bent shaft, the following features were selected:

1. Kurtosis of the time synchronous average of the output signal vibration.
2. Ratio between peak at 1xrpm of input shaft to next highest peak after wavelet decomposition and FFT of approximation signal.

The wavelet decomposition was used to isolate both the impact and the 1X harmonic component of the input shaft (5X of output shaft); including indicators that quantify the impact and the harmonic component is necessary for the bent shaft diagnosis.

Figure 19, provides an example from a regime (speed= 50 Hz, Load=Light, Gear=Helical) where the probability of defect was calculated, and it is clear that four of these data files can be isolated from the other files because they have a higher probability of defect for a bent input shaft.

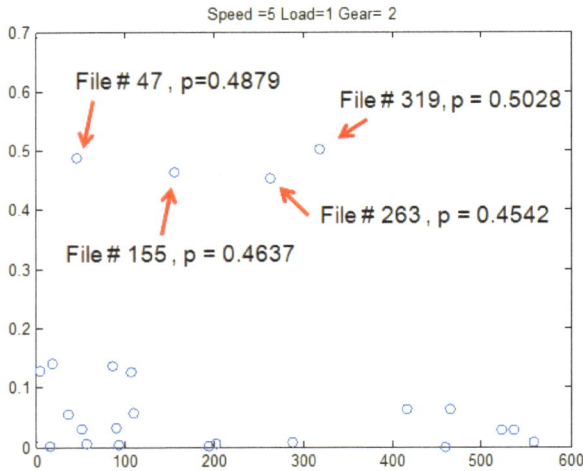

Figure 19: Probability of input bent shaft defect for regime of 50Hz speed, light load, and helical gears.

6.5 Bearing Fault Detection

For detecting the inner race defect in the bearing on the input shaft output side, the feature selected was the ratio of feature 1 below divided by feature two:

1. Peak at Ball Pass Frequency Inner Race (BPFI) from the Envelope spectrum of the output accelerometer, input shaft bearing.
2. Peak in time synchronous average spectrum at input shaft speed from input and output accelerometer.

Figure 20, shows an example from a regime (speed= 50 Hz, Load=Light, Gear=Helical) where the probability of defect was calculated, it is clear that four of these data files can be isolated from the other files because they have a higher probability of inner race defect for the bearing on the input shaft/output side.

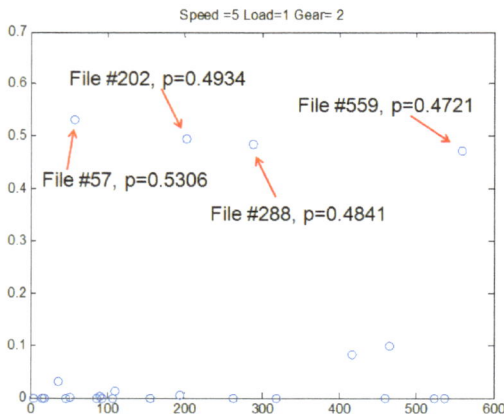

Figure 20: Probability of inner race bearing defect on the input shaft/output side bearing calculated for regime of 50Hz speed, light load, and helical gears.

6.6 Detecting Bad Key Fault

For this gearbox system, the output shaft could be in two different states, either the health state or a defect state; where the defect state is due to a loose coupling between the output shaft and the load because of a faulty key. This faulty shaft key would cause slipping between the output shaft and the load; the vibration signals and the tachometer information can be used to isolate this particular event.

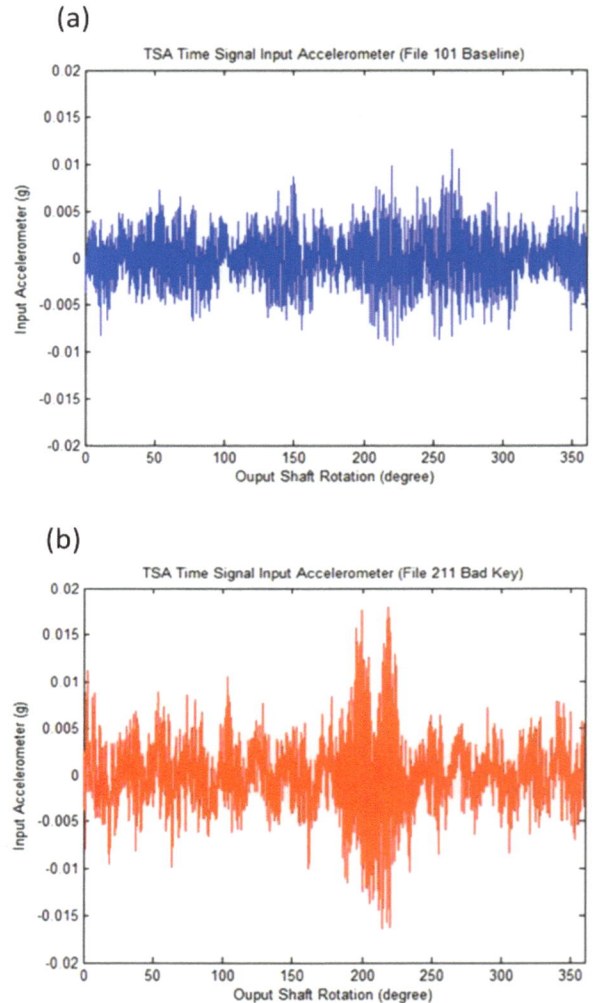

(a)

(b)

Figure 21: Signature of bad key defect.

Figure 21 (b) shows the time synchronous average signal of a data set with a bad key. It can be noticed from the signal of an impact that is of much longer duration; this is due to the averaging of the time synchronous signal and the slipping of the shaft speed due to the fault key. Notice that this signature of a bad key is in sharp contrast to the time synchronous signal shown in Figure 21 (a) for a gearbox without any faults including a key that is working properly. Also if one would compare the signature of a broken tooth compared to a bad key by examining the time synchronous average time signal; although in both cases there is this transient impact, the bad key has a specific impact that last for a much longer duration due to the slipping of the output shaft. Utilizing a specific set of features that can isolate the bad key from not only the baseline case, but also faults that also have transient impacts is what is needed for providing the specific root cause diagnosis information.

For detecting the bad key defect on the output shaft, the following features were selected:

1. Kurtosis from time synchronous average time signal from output accelerometer.
2. Spectral Kurtosis feature from output accelerometer, for band from 10 KHz to 20 KHz.

Figure 22 shows the probability of defect results for the bad key case; the result show 4 data sets in a particular regime that clearly have this problem.

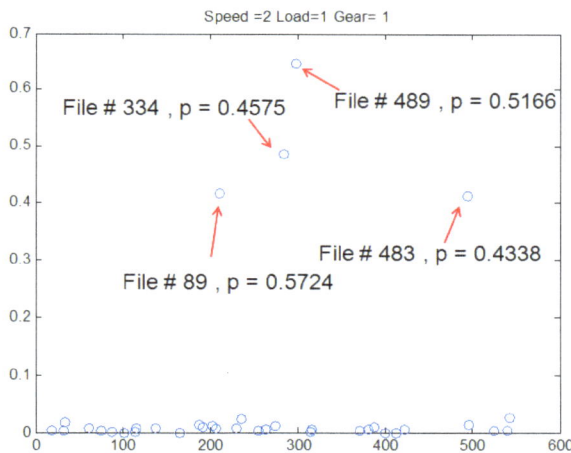

Figure 22: Probability of bad key defect for regime of 35Hz speed, light load, and spur gears.

7. CONCLUSION

This paper introduced a systematic methodology for gearbox health assessment and fault classification. The methodology was validated for 560 data sets of gearbox vibration data provided by the Prognostics and Health Management Society for the 2009 data challenge competition, and won the first place in the student division. The methodology involves the utilization of a comprehensive set of signal processing and feature extraction methods; in that the use of a single signal processing method would not be applicable for a mechanical system that consisted of a multitude of different faults. A regime segmentation approach was

necessary to provide a fair comparison between data sets and grouped the data by load, speed, and gear type. A health assessment algorithm was used to classify and find the 80 baseline healthy data sets. Using the baseline data sets provided by the health assessment method, a fault diagnosis method based on a modified Euclidean distance calculation from normal along with specific features correlated to different fault signatures is used to diagnose specific faults. The fault diagnosis method is evaluated for the diagnosis of five different gearbox fault types, and could be further extended for other faults as long as a set of features can be correlated with a known fault signature. The methodology can be further applied to other rotating machine applications involving gear, shaft, or bearing components

8. SUGGESTIONS FOR FUTURE WORK

Some of the future work looks to further refine some of the techniques and methods employed as follows:

1. Refine the distance calculation algorithm used for fault diagnosis. The current modified Euclidean distance function used in this paper has proven to be very useful for vibration-based diagnosis applications and can be further enhanced to take baseline variation into consideration whenever such data is available

2. Additional signal processing methods such as the empirical mode decomposition could be evaluated to see if a more robust feature set can be provided for gearbox mechanical defects.

3. A "regime-independent" fault signature discovery method would be evaluated. Such a method would be very useful for diagnosis; in such that whenever a specific fault is identified in one regime, the signature could be captured and used to find similar faults in other regimes.

REFERENCES

J. Antoni, and R.B. Randall (2006). The spectral kurtosis: application to the vibratory surveillance and diagnostics of rotating machines, *Mechanical Systems and Signal Processing*, vol. 20, pp. 308-331.

J.Antoni. (2006). The spectral kurtosis: a useful tool for characterizing nonstationary signals, *Mechanical Systems and Signal Processing*, vol. 20, pp. 282-307.

E. Bechhoefer, and A. Bearnhard (2003). Setting HUMS Condition Indicator Thresholds by

Modeling Aircraft and Torque Band Variance, *IEEE Aerospace Conference Proceedings*.

F.K. Choy, S. Huang, J.J. Zakrajsek, R.F. Handschuh, and D.P. Townsend (2004). Vibration signature analysis of a faulted gear transmission system, *NASA Technical Memorandum*, NASA TM-106623/ARL-TR 475/AIAA-94-2937.

H.J. Decker, and D.G. Lewicki. (2003). Spiral Bevel Pinion Crack Detection in a Helicopter Gearbox, *ARL-TR-2958*, U.S. Army Research laboratory, NASA Lewis.

P. Grabill, J. Berry, L. Grant, and J. Porter. (2001). Automated Helicopter Vibration Diagnosistcs for the US Army and National Guard, *57th Annual Forum of the American Helicopter Society*, Washington, DC, pp.1831-1842.

J. Keller and P. Grabill. (2003). Vibration Monitoring of UH-60A Main Transmission Planetary Carrier Fault, *The American Helicopter Society 59th Annual Forum*, Phoenix, Arizona.

B. Li, M. Chow, Y. Tipsuwan, and J. Hung (2000). Neural-Network Based Motor Rolling Bearing Fault Diagnosis, *IEEE Transactions on Industrial Electronics*, vol. 47, pp. 1060-1069.

J. Ma. (1995). Energy Operator and Other Demodulation Approaches to Gear Defect Detection, *Proceedings of the 49th Meeting of the Society for Machinery Failure Prevention Technology*, Vibration Institute, Willobrook, Illinois, pp.127-140.

Z.K. Peng, and F.L. Chu. (2003). Application of the wavelet transform in machine condition monitoring and fault diagnostics: a review with bibliography, *Mechanical Systems and Signal Processing*, vol. 18, pp. 199-221.

PHM 2009 Data challenge Competition, (2009). [http://www.phmsociety.org/competition/09]

PHM Society Public Datasets, (2009).PHM 2009 Data challenge Competition, Labeled Dataset. [https://www.phmsociety.org/references/datasets]

P.D. Samuel and D.J. Pines. (2005). A review of vibration-based techniques for helicopter transmission diagnostics, *Journal of Sound and Vibration*, vol. 282, pp. 475-508.

J. Sheeley, R. Xu, Z. Ren, B. Ayhan, W. Lee, M.S. Sahni, Hu Qiaohui, and T. McClerran (2009). Initial Operational Evaluation of a Novel Corona Monitoring System, *Proceedings of the Society for Machinery Failure Prevention Technology (MFPT)*.

P.W. Tse, Y.H. Peng, and R. Yam (2001). Wavelet Analysis and Envelope Detection for Rolling Element Bearing Fault Diagnosis-Their Effectiveness and Flexibilities, *Journal of Vibration and Acoustics*, vol.123, pp.303-310.

P. Vecer, M. Kreidl, and R. Smid (2005). Condition Indicators for Gearbox Condition Monitoring Systems, *Acata Polytechnica*, vol. 45, No. 6

H. Zheng, Z. Li, and X. Chen. (2002). Gear fault diagnosis based on the continuous wavelet transform, *Mechanical Systems and Signal Processing*, vol.16, pp. 447-457.

Bearing fault detection with application to PHM Data Challenge

Pavle Boškoski [1], and Anton Urevc [2]

[1] *Jožef Stefan Institute, Ljubljana, Slovenia*
pavle.boskoski@ijs.si
[2] *Centre for Tribology and Technical Diagnostics,*
Faculty of Mechanical Engineering,
University of Ljubljana, Slovenia
anton.urevc@ctd.uni-lj.si

ABSTRACT

Mechanical faults in the items of equipment can result in partial or total breakdown, destruction and even catastrophes. By implementation of an adequate fault detection system the risk of unexpected failures can be reduced. Traditionally, fault detection process is done by comparing the feature sets acquired in the faulty state with the ones acquired in the fault–free state. However, such historical data are rarely available. In such cases, the fault detection process is performed by examining whether a particular pre–modeled fault signature can be matched within the signals acquired from the monitored machine. In this paper we propose a solution to a problem of fault detection without any prior data, presented at PHM'09 Data Challenge. The solution is based on a two step algorithm. The first step, based on the spectral kurtosis method, is used to determine whether a particular experimental run is likely to contain a faulty element. In case of a positive decision, fault isolation procedure is applied as the second step. The fault isolation procedure was based on envelope analysis of band–pass filtered vibration signals. The band–pass filtering of the vibration signals was performed in the frequency band that maximizes the spectral kurtosis. The effectiveness of the proposed approach was evaluated for bearing fault detection, on the vibration data obtained from the PHM'09 Data Challenge.

1. INTRODUCTION

Stable and predictable condition of process equipment, high process availability and reliability are key factors that keep a company competitive. However, wear, material stress and environmental influences can cause mechanical faults which result in equipment breakdowns. Since the emergence of faults is inevitable, it is of utmost importance to construct an effective fault detection system capable of detecting these faults in their incipient phase. Such early detection will prevent unscheduled interruptions in the machine operation, which effectively will increase the overall performance.

In the research domain there is an impressive body of literature that addresses the issues of fault detection. One group of authors mainly focuses on modeling the vibration signals generated by a specific mechanical element, like gears, bearings etc. In that manner, (Bartelmus, 2001) developed model for vibrations produced by meshing gears. (Endo, Randall, & Gosselin, 2009) developed models for gear vibrations under specific tooth faults. Similarly, (Tandon & Choudhury, 1999) derived the relations for the principle frequencies components in the vibration signals produced by localized bearing faults. Although the fault signatures have been thoroughly examined, the detection of the faults in their incipient phase has proved to be a difficult task. Many authors have addressed this issue by employing variety of signal processing techniques. The envelope spectrum analysis of the vibration signals has been one of the most commonly used methods (Rubini & Meneghetti, 2001; Ho & Randall, 2000). However, several authors have shown that a significant increase in the sensitivity of the envelope analysis can be achieved by calculating the envelope spectrum of a filtered signal. In that manner (Wang, 2001) proposed filtering the vibration signals around the system's resonance frequency, (Staszewski, 1998) used wavelet denoising techniques, (Sawalhi, Randall, & Endo, 2007) used spectral kurtosis method for determining the most suitable frequency band.

The majority of the listed methods have been successfully applied in systems where historical data of the fault–free state have been previously acquired, so that any small deviation from the fault–free case could be detected. Conversely, such historical data of the fault–free system were not available on the set of signals acquired from the PHM-09 test–rig (PHM,

2009). In absence of a base–line determining the fault–free runs, the fault detection procedure was implemented using a two–step approach. The first step is used to decide whether the observed experimental run is likely to belong to a faulty machine or not. Provided the observed run is associated to the faulty condition, the second step is applied. This step consists of a fault diagnosis process, in which the fault origins are determined.

The decision whether a machine state is faulty or fault–free was based on the maximal value of the spectral kurtosis (SK) of the acquired vibration signals for a particular run. Such a choice is based on the property of SK for detection of transients, which is explained in more details in Section 4. Additional results of the SK calculation process are the frequency band parameters, central frequency f_c and bandwidth B_w, where the SK maximum resides. These parameters are used to filter the vibration signals. The filtered signals are then used to calculate the envelope spectrum. The resulting spectrum is afterward used as a starting point for the fault detection procedure. Despite the lack of data from the fault–free motor run, the proposed two–step fault detection approach has proved capable of detecting bearing faults on the PHM'09 test–rig (PHM, 2009).

The presentation of the proposed algorithm in this paper is organized as follows. In Section 2 we present the basics of the bearing fault frequency and the used bearing model. A brief overview of the envelope analysis and spectral kurtosis methods are given in Section 3. A detailed description of the proposed fault detection algorithm, with a proposed simplification procedure are presented in Section 4.

2. BEARING FAULTS

In a presence of localized bearing fault, impacts occur every time bearing's roller element passes over the damaged area. Each of these impacts excites an impulse response of the observed bearing, i.e. exponentially decaying oscillation $s(t)$. Under the assumption that the bearing rotates with constant rotational frequency f_{rot}, these impulses will be periodic with some period T. The period T is directly connected to the type of the localized fault. In can be considered that $T = 1/f_e$, where f_e is one of the principle bearing fault frequencies (Tandon & Choudhury, 1999):

$$
\begin{aligned}
f_{BPFO} &= \frac{Z f_{rot}}{2}\left(1 - \frac{d}{D}cos\alpha\right) \\
f_{BPFI} &= \frac{Z f_{rot}}{2}\left(1 + \frac{d}{D}cos\alpha\right) \\
f_{FTF} &= \frac{f_{rot}}{2}\left(1 - \frac{d}{D}cos\alpha\right) \\
f_{BSF} &= \frac{D f_{rot}}{2d}\left(1 - \left(\frac{d}{D}cos\alpha\right)^2\right),
\end{aligned}
\tag{1}
$$

where f_{BPFO} is the ball passing frequency of the outer race, f_{BPFI} is the ball passing frequency of the inner race, f_{FTF} is the fundamental train frequency and f_{BSF} is the ball spin frequency, Z is the number of rolling elements, d is the rolling element diameter, D is the pitch diameter, α is the contact angle and f_{rot} is the inner ring rotational speed.

The vibrations $x(t)$ produced by a localized bearing fault may be written as

$$
x(t) = \sum_{i=-\infty}^{+\infty} s(t - iT). \tag{2}
$$

Although the rotation frequency f_{rot} may be considered as constant, small speed fluctuations are always present. Additionally the speed of the roller element which enters the load zone slightly differs from the one that is outside the load zone. Such random fluctuations are expressed as a time lag of occurrence of each impact, in particular

$$
x(t) = \sum_{i=-\infty}^{+\infty} s(t - iT - \tau_i), \tag{3}
$$

where τ_i represents the time lag of occurrence of the ith impact.

In Eq. (3) we considered that all impacts have same amplitude. However, each impact excites an impulse with somewhat different amplitude due to changes in the bearing surface, the way the ball enters the damaged region etc. Such random changes are incorporated by adding a factor A_i, which represents the amplitude of ith impact

$$
x(t) = \sum_{i=-\infty}^{+\infty} A_i s(t - iT - \tau_i). \tag{4}
$$

Finally, in order to take into account all surrounding non–modeled vibrations as well as any other environmental influence, a purely random component $n(t)$ is added to Eq. (4). The final model of bearing vibrations becomes (Randall, Antoni, & Chobsaard, 2001):

$$
x(t) = \sum_{i=-\infty}^{+\infty} A_i s(t - iT - \tau_i) + n(t). \tag{5}
$$

3. METHODS OVERVIEW

The vibration signal defined by (5) is an amplitude modulated (AM) signal, where the modulation itself can be considered as a random signal. The information about the present fault is contained in the mean period T of the occurrence of the impacts. Hence all the needed diagnostic information is contained within the signal's envelope. When the amplitudes A_i in (5) are high enough a fairly simple analysis of the envelope spectrum is sufficient for the fault diagnosis process. However, when these amplitudes are small and dominated by the noise $n(t)$, the simple envelope spectrum analysis turns to be

---

In the case of PHM'09 challenge all bearings were ball bearings

ineffective in extracting information about the fault. A way around this problem is to calculate the envelope of the signal after the vibration signal has been let through the band–pass filter centered at a selected carrier frequency. An improper selection of band–pass filtering parameters can significantly hinder the effectiveness of the fault detection process. As a result of this effect, a variety of different approaches have been developed offering different solutions to the issue of band–pass filter parameter selection. In our case we adopted the spectral kurtosis method, which has proved capable of detection transients buried in noise.

3.1 Spectral Kurtosis

The spectral kurtosis method was firstly introduced by (Dwyer, 1983), as a method that is able to distinguish between transients (impulses and unsteady harmonic components) and stationary sinusoidal signals in background Gaussian noise. Spectral kurtosis takes high values for frequency bands where the vibration signal $x(t)$, defined with Eq.(5), is dominated by the corresponding impulses, and it takes low values for frequency bands where the signal is dominated by the Gaussian noise $n(t)$ or stationary periodic components. If we rewrite the signal from Eq.(5) as

$$x(t) = y(t) + n(t), \tag{6}$$

where

$$y(t) = \sum_i A_i s(t - iT - \tau_i), \tag{7}$$

than the SK values $K_x(f)$ for the signal $x(t)$ contaminated by additive noise $n(t)$ can be calculated as (Antoni & Randall, 2006)

$$K_x(f) = \frac{K_y(f)}{[1 + \rho(f)]^2}, \tag{8}$$

where $K_y(f)$ is the spectral kurtosis of the signal $y(t)$, and $\rho(f)$ is the noise–to–signal ratio for that particular frequency f. The value for $K_y(f)$ can be obtained using the following relation

$$K_y(f) = \frac{S_{4y}(f) - 2S_{2y}^2(f)}{S_{2y}^2(f)}, \tag{9}$$

where $S_{2y}(f)$ and $S_{4y}(f)$ are the second and fourth spectral moments respectively. The maximum of Eq.(8), actually determines the frequency band where the signal–to–noise ratio in the observed signal is the biggest and at the same time the observed signal $x(t)$ is the closest to the original, uncontaminated signal, $y(t)$.

The definition of SK given by the Eq.(9) bears resemblance with the statistical definition of kurtosis. However the actual physical interpretation and its ability for detection of non-stationary transients in signals is not so obvious. One way for clarification is to observe the time–frequency characteristic of the simulated vibration signal $x(t)$, defined by Eq. (5). The simulation was conducted with $T = 333$ Hz, $s(t) =$

$e^{-100t} \sin(2\pi 1500t)$ and SNR=1. The random time fluctuations τ_i and random amplitudes A_i were Gaussian random processes with zero–mean with $\sigma_{\tau_i} = 0.05T$ and $\sigma_{A_i} = 0.5$ respectively (cf. Figure 1).

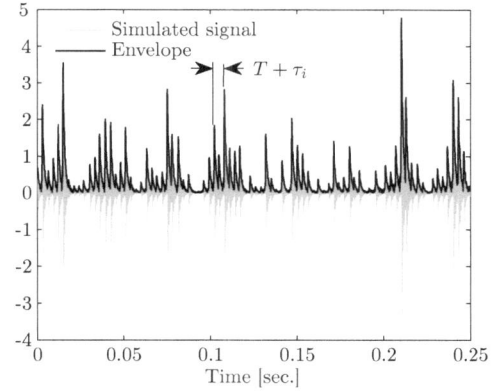

Figure 1. Simulated signal $x(t)$, Eq. (5)

The time–frequency characteristic of the simulated signal is shown in Figure 2. It is noticeable that the highest peaks are around 1.5 kHz, which was the chosen simulated eigenfrequency of the impulse response $s(t)$. The amplitudes of the spectral components around 1.5 kHz vary in time considerably, compared to the ones above and below this frequency band. By calculating the average over time we will obtain the standard power–spectral density (PSD).

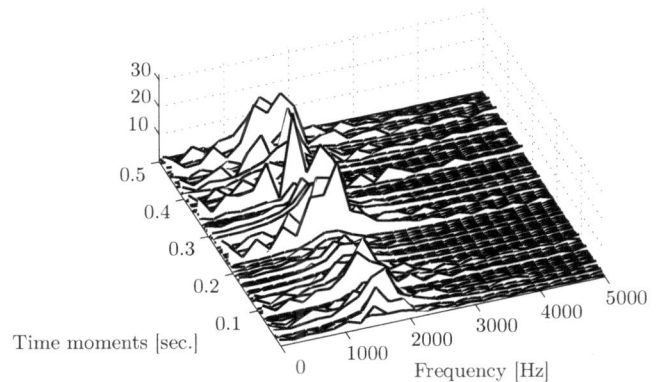

Figure 2. Time–frequency characteristic of the simulated signal $x(t)$ Eq. (5)

If we now consider the changes of the amplitude of particular spectral components as a stochastic process, the spectral kurtosis actually searches for the frequency band where this stochastic process exhibits the highest kurtosis. For our simulated signal Eq. (5), the frequency band in question is around the resonance frequency of 1.5 kHz. The time change of the amplitude of that spectral component compared with its average value is shown in Figure 3.

It can be concluded that for non–stationary processes these

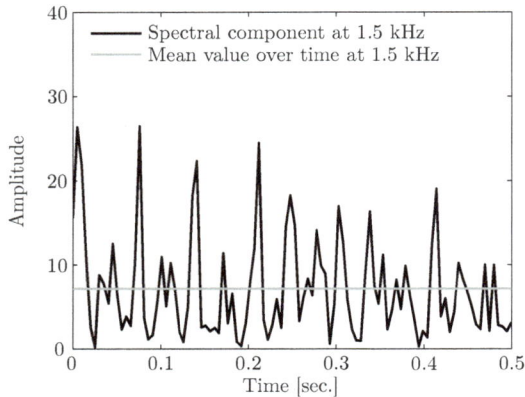

Figure 3. Changes of amplitude of the spectral component at 1.5 kHz over time

discrepancies in amplitude of some spectral components in time will be more expressed then in the cases of stationary processes. Consequently, we can use the SK as an indicator for a frequency band where the signal's non–stationarities are most expressed.

The estimation of the time–frequency characteristics of the simulated signal was based on the Short–time Fourier transform (STFT). Since the resolution in time and in frequency depends on the used window length, the search for the best frequency range is performed by examining the SK value for several window lengths (Antoni, 2007). This procedure produces a diagram called *kurtogram*. The kurtogram diagram for the simulated signal $x(t)$ defined by Eq. (5) is shown in Figure 4.

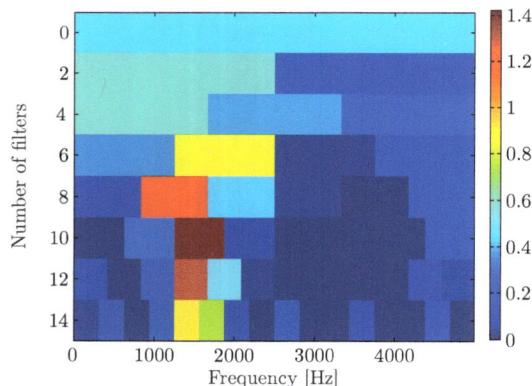

Figure 4. Kurtogram of the simulated signal $x(t)$ defined by Eq. (5)

From the kurtogram shown in Figure 4, we can see that the maximal value of SK is obtained in the frequency band with central frequency $f_c = 1500$ Hz and bandwidth $B_w = 625$ Hz, and the maximum of SK in that frequency band is 1.4.

For a comprehensive derivation and all properties of SK one

should refer the following references (Antoni, 2006; Antoni & Randall, 2006; Sawalhi et al., 2007).

3.2 Envelope analysis

After filtering the signal in the frequency range determined by the SK method, the final step in the fault detection process is done by envelope analysis. The use of envelope analysis is justified, because the information about the fault is in the occurrence period T of the quasi–periodic impact impulses. For the simulated signal $x(t)$ defined by Eq. (5), the envelope is marked with black line in Figure 1.

The signal's envelope is obtained from the Hilbert transform, i.e. by analyzing the amplitude of the analytical signal $x_a(t)$. The analytical signal $x_a(t)$ is a complex signal whose real part is the original signal $x(t)$, and the imaginary part is the Hilbert transform of the original signal $x(t)$

$$x_a(t) = x(t) + i\mathscr{H}[x(t)], \qquad (10)$$

where $\mathscr{H}[x(t)]$ is the Hilbert transform of the signal $x(t)$

$$\mathscr{H}[x(t)] = \frac{1}{2\pi}\int_{-\infty}^{+\infty}\frac{x(t)}{t-\tau}\,d\tau. \qquad (11)$$

Fourier transform of the analytical signal $x_a(t)$ is:

$$
\begin{aligned}
X_a(t)(f) &= X(f) + j\mathscr{H}[X(t)] \\
&= \begin{cases} 2X(t) & \text{for } f > 0, \\ X(t) & \text{for } f = 0, \\ 0 & \text{for } f < 0. \end{cases}
\end{aligned} \qquad (12)
$$

This shows that the analytical signal has spectrum only in the positive frequency range. By calculating the amplitude of the analytical signal (10)

$$a(t) = \sqrt{x^2(t) + \mathscr{H}^2[x(t)]} \qquad (13)$$

we obtain the envelope of the signal.

The spectrum of the envelope of the simulated signal $x(t)$ is shown in Figure 5. From that spectrum we can easily identify the spectral component at 333 Hz, which is the period of the occurrence of the simulated impacts.

3.3 Alternative way of estimating signal envelope

Although the calculation of the signals envelope using the Hilbert transform can be performed using Fast Fourier transform (FFT), a good estimation can be done by calculating the spectrum of the absolute value of the signal (Benko, Petrovčič, Musizza, & Juričić, 2008). By doing so we skip one FFT calculation of the whole signal. Namely, calculating envelope spectrum using the Hilbert transform we first calculate FFT of the original signal. Then we obtain the analytical signal $x_a(t)$ by employing the Eq. (12). As the last step we calculate FFT of $x_a(t)$, thus obtaining the envelope spectrum. Using the estimation method for envelope spectrum, we just have to calculate one FFT of the $|x(t)|$.

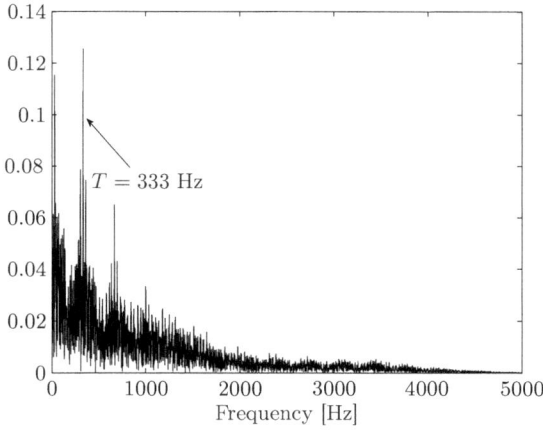

Figure 5. Envelope spectrum of the simulated signal $x(t)$, defined by Eq. (5)

This procedure will be presented by analyzing the envelope of an amplitude modulated (AM) signal $y_{am}(t)$ in form:

$$y_{am}(t) = [C + m(t)]cos(\omega_c t) \\ = y_0(t)cos(\omega_c t), \tag{14}$$

where $y_0(t) = [C + m(t)]$ is the modulation signal, or the envelope, and $\omega_c = 2\pi/T_c$ is the carrier frequency.

The Fourier transform of the absolute value $|y_{am}(t)|$ can be obtained as

$$\mathcal{F}\{|y_{am}(t)|\} = \sum_p \int_{T_p} y_{am}(t)e^{-j\omega t}dt \\ - \sum_q \int_{T_q} y_{am}(t)e^{-j\omega t}dt \\ = \mathcal{F}\{y_{am}(t)\}(\omega) \cdot \\ - 2\sum_q \int_{T_q} y_{am}(t)e^{-j\omega t}dt, \tag{15}$$

where T_p is one of the intervals where the signal $y(t)$ is positive, and T_q is one of the intervals where $y(t)$ is negative. In lower frequencies $\omega_0 < \omega_{max}$, the spectrum of the observed signal is zero, i.e. $\mathcal{F}\{y_o(t)\}(\omega) = 0, \forall \omega \geq \omega_{max}$, where $0 < \omega_{max} \ll \omega_c$. Under these conditions the Eq. (15) can be rewritten as

$$\mathcal{F}\{|y_{am}(t)|\}(\omega_0) = \\ = -2\sum_q \int_{T_q} y_{am}(t)e^{-j\omega_0 t}dt \\ = -2\sum_q \int_{T_q} [C + m(t)]cos(\omega_c t)e^{-j\omega_0 t}dt. \tag{16}$$

If we now allow $\omega_c \to \infty$, the intervals T_q will tend to zero, and within the observed interval the function $[C + m(t)]cos(\omega_c t)e^{-j\omega_0 t}$ may be considered as constant. The observed value is the value of the function in moment t_q, which

is the middle of the T_q interval. Under these assumptions Eq. (16) becomes

$$\lim_{\omega_c \to \infty} \mathcal{F}\{|y_{am}(t)|\}(\omega_0) = \\ = -2\sum_q [C + m(t_n)]e^{-j\omega_0 t_n} \int_{T_q} cos(\omega_c t)dt = \\ = -2\sum_q [C + m(t_n)]e^{-j\omega_0 t_n} \left(-\frac{2}{\omega_c}\right) = \\ = 2\sum_q [C + m(t_n)]e^{-j\omega_0 t_n} \frac{T_c}{\pi}. \tag{17}$$

Since we have allowed $\omega_c \to \infty$, consequently the period $T_c \to 0$. In such case we can change the summation to integration and the Eq. (17) becomes

$$\lim_{T_c \to 0} \mathcal{F}\{|y_{am}(t)|\}(\omega_0) = \frac{2}{\pi} \int [C + m(t)]e^{-j\omega_0 t}dt. \tag{18}$$

The last part of Eq. (18) is actually the Fourier transform of the signals envelope. Finally we can conclude that when $\omega_0 \ll \omega_c$ then

$$\mathcal{F}\{|y_{am}(t)|\}(\omega_0) \propto \mathcal{F}\{C + m(t)\}(\omega_0). \tag{19}$$

4. FAULT DETECTION PROCEDURE

Traditionally, the presence of fault is determined based on a results of a spectrum comparison with previous data (fault–free data) (Sawalhi & Randall, 2008). Since such set of fault–free data was unavailable, we had to use a different approach in order to detect the faulty experimental runs. We have adopted a two step approach. Firstly we have constructed a set of experimental runs that are most likely to represent a run with faulty element. As a second step, each of these presumably faulty runs was analyzed using envelope spectra of the filtered signal in the frequency band that maximizes the SK value.

Since the PHM'09 Data challenge consisted of 560 test–runs conducted under 5 different running speeds, we have divided the complete data set into 5 batches each comprising experiments conducted under the same rotational speed f_{rot}. Such a pre–processing step was necessary since the principle bearing fault frequencies, defined by Eq. (1), are dependent on rotational speed f_{rot}. Thus we have analyzed 5 different batches consisting of 112 runs each using the same algorithm. The results presented in this section refer to only one group, but the same principle applies to the remaining 4 batches.

The decision, whether a particular experimental run is likely to contain a faulty element or not, was based on the maximum value of the SK for the particular measurement. The idea behind this approach is the property of the SK which states (Antoni & Randall, 2006) that the value of SK $K_x(f)$, defined by Eq. (8), increases with the intensity of the fluctuations in the impulse amplitudes. Consequently the value of the SK can be used as an indication of the severity of the damage. So SK value was calculated for each measurement separately and the measurements were sorted by decreasing values of SK. Thus, the measurements at the top of the list were more likely to represent faulty runs, and those at the bottom of the list were considered as more likely to be fault–free runs.

In the second step, the fault isolation procedure, started by analyzing measurements from the top of the sorted list. Each signal from the list was filtered in the frequency band in which the value of SK was the highest. After that the envelope spectrum of such filtered signal was calculated by applying the Eq. (19), although one might also opt for the calculation of the signal's envelope based on Hilbert transform and Eq. (13). The fault detection was based on the amplitudes of spectral components at the bearing characteristic frequencies, defined by Eq. (1), obtained from the envelope spectrum. A block diagram of the described approach is shown in Figure 6.

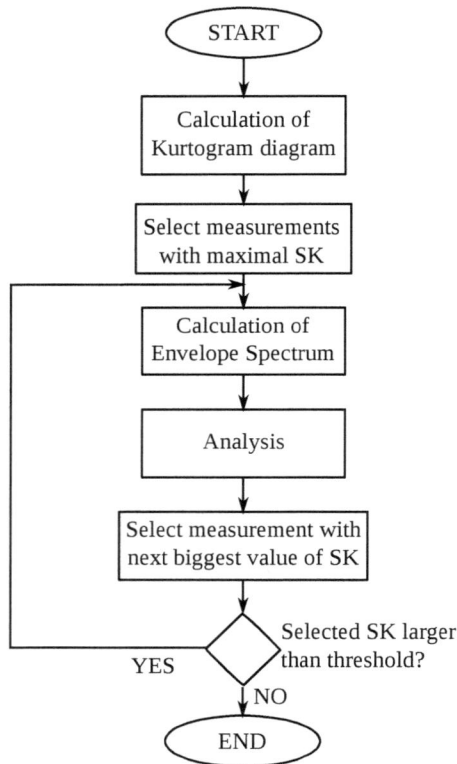

Figure 6. Block diagram of the used fault detection algorithm

After the analysis of all experimental runs from the top of the list, the next step is to determine whether the remaining measurements consist of only fault–free runs. In this is the case the algorithm ends. The decision whether the remaining runs contain experiments with damaged elements is based on the value of SK. The values of SK for a group of 112 measurements for one particular speed is shown in Figure 7. Under assumption that the fault–free runs have the smallest value of

SK, we can determine a threshold that can serve as a boundary between faulty and fault–free runs. The procedure for selection of the threshold takes into account the fact that SK values for multiple measurements with same kind of fault should have similar values. Therefore, if we take a window consisting of several neighboring measurements and calculate the interquartile interval on their SK values, we can use the width of the interquartile interval to determine whether the measurements originate from the same machine configuration or not. If they do originate from the same configuration, the width of the interquartile interval will be smaller. Otherwise it will be significantly larger. Thus, we can use the width of interquartile interval as an indication which neighboring measurements have been done in different machine configuration.

The first change in the machine configuration occurs when some kind of fault was introduced into a fault–free machine. In order to identify this change we analyzed the interquartile intervals of the measurements with lowest value of SK. The values of the interquartile intervals are shown in Figure 8 using a box plot. From the figure we can detect that the measurements with the smallest width of the interquartile interval are found for measurements 18,19 and 20. This leads to a conclusion that the first change in the machine configuration occurred several measurements before i.e. somewhere between the 14th and the 17th measurement. In this interval the 16th measurement has the widest interquartile interval. Therefore, as a threshold we have chosen the upper interquartile value of the 16th measurement.

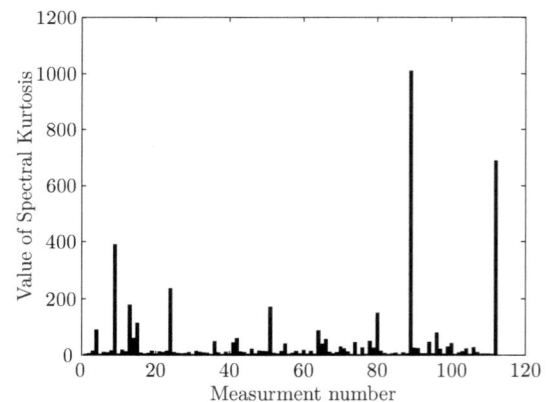

Figure 7. Values of the spectral kurtosis for 112 measurements

Although we have decided that 16 measurements with the lowest value of SK represent fault–free runs, it can be noticed that there are two segments of measurements with smaller interquartile intervals within these 16 measurements, in particular the 3rd, 4th and the 5th measurement, as well as the 10th and the 11th. These two intervals indicate the two different load levels under which the machine was operating during the conducted experiments. However these changes did not influence

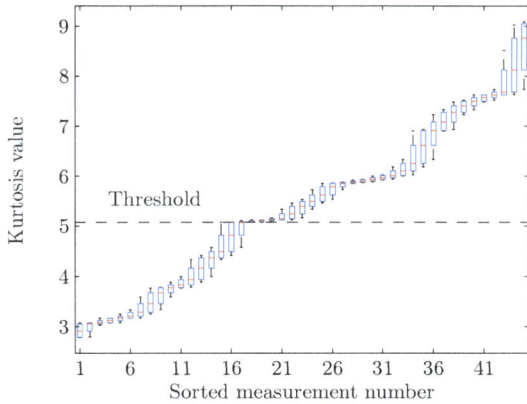

Figure 8. Box plot for the values of SK for determining the threshold (only the lower values of SK are depicted)

the overall fault detection process.

4.1 Results

The values of SK for a batch of 112 measurements with same speed is shown in Figure 7. We can notice that some measurements show high values of SK compared to the rest of the batch. As already explained, the iterative procedure starts with the measurement that has the highest SK.

The kurtogram diagram for the experimental run with the highest kurtosis is shown in Figure 9. In this particular case, the frequency band with the highest SK can be obtained by filtering the original signal with a band–pass filter with central frequency $f_c = 18.7$ kHz and filter bandwidth $B_w = 1.4$ kHz.

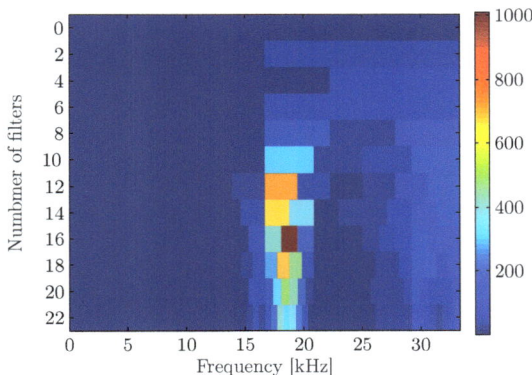

Figure 9. Kurtogram for the measurement from the top of the sorted list

By using this filter parameters we calculated the envelope spectrum of the examined measurement, shown in Figure 10. The envelope spectrum is dominated by a single spectral component at f_{BPFI}.

Since the spectrum is dominated by a single spectral component the fault diagnosis procedure is fairly simple. However,

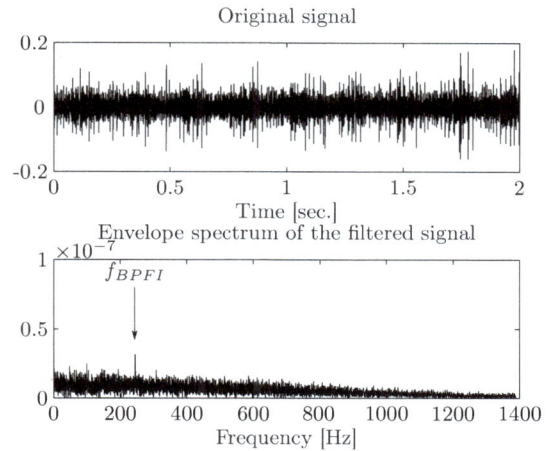

Figure 10. Envelope spectrum of the filtered vibration signal from the output shaft sensor

the final task for the data challenge was the selection which of the 6 bearings was damaged. All three shafts of the gear box rotated with different rotating speeds f_{rot}. Since all 6 bearings were of the same type, each pair of bearings on a particular shaft had its own set of bearing fault frequencies defined by Eq. (1). So, as a first step, we have located the shaft on which the faulty bearing was mounted. Afterwards, the selection which of the two possible bearings was damaged was done by observing the values of the signals acquired at both ends of the gearbox. The hypothesis is that the closer the sensor is to the fault itself the larger are the amplitudes of the corresponding bearing fault frequency. For the examined case of bearing inner race fault, the envelope spectrum of the filtered signal from the sensor mounted on the input shaft is shown in Figure 11. It is visible that the spectrum contains the spectral component at f_{BPFI}, however the amplitude is 50 times smaller than the corresponding one from the sensor mounted on the output shaft, shown in Figure 10. So for this particular experimental run we can conclude that the fault was on the inner race of the bearing on the output side of the gearbox.

The task of fault detection becomes more complicated as the severity of the fault decreases. One such case is bearing with damaged ball. This fault is characterized with increased amplitude of a spectral component at f_{BSF} (cf. Eq. 1). The kurtogram diagram for this fault is shown in Figure 12. Unlike the diagram for bearing inner race fault, the maximum value of the SK for this case is 4.5, compared to the SK value of 1000 observed in the previous case. Another important observation is that the impulses generated by this fault are most visible around $f_c = 7.5$ kHz. Hereupon, we can conclude that different faults may excite different eigenfrequencies of the observed system. So, for each run we have to calculate the maximal value of SK, thus obtaining the proper frequency band parameters valid for that particular run. Calculated enve-

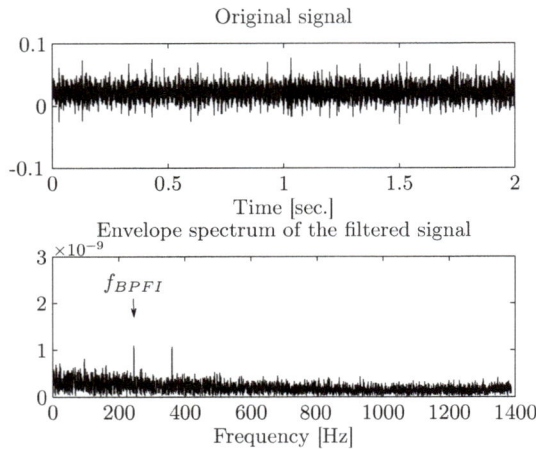

Figure 11. Envelope spectrum of the filtered vibration signal from the input shaft sensor

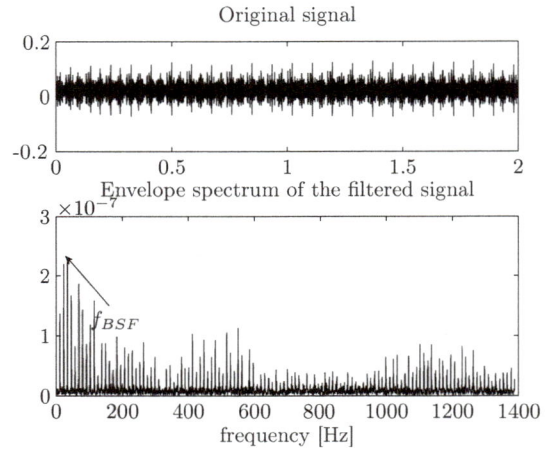

Figure 13. Envelope spectrum of the filtered vibration signal from the input shaft sensor

lope spectrum of the filtered signal around the proposed central frequency f_c is shown in Figure 13. Similarly to the first case, the fault detection for the currently observed run is quite straightforward. The most dominant spectral component is centered at f_{BSF}, which unambiguously points towards roller element damage.

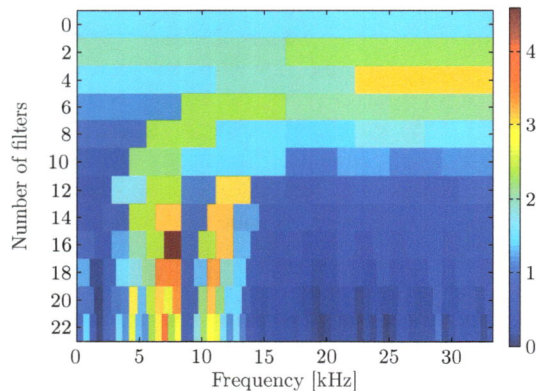

Figure 12. Kurtogram for the measurement from the top of the sorted list

4.2 Procedure simplification for cases with heavy damage

Spectral kurtosis is quite efficient for cases when the transient components in the measured signals have small energy compared to other vibration sources in the vicinity, like meshing gears or any other environmental influence. However, for the cases where the amplitude of impulses produced by the observed fault was sufficiently high, the fault detection procedure can be done by calculating the envelope spectrum of the whole unfiltered signal. Such a spectrum for the case of f_{BPFI} fault is shown in Figure 14. Comparing this spectrum

with the one shown in Figure 10, we can conclude that both spectra are dominated by the same frequency component.

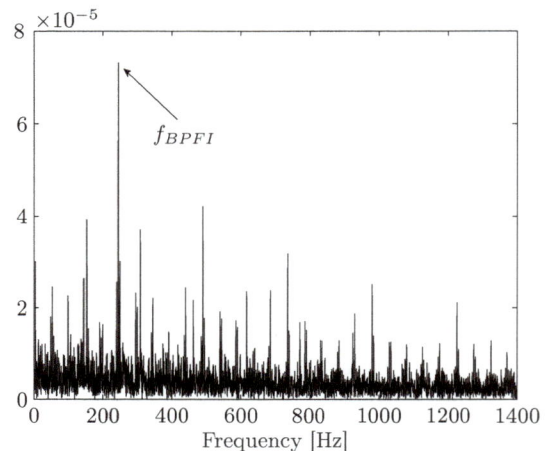

Figure 14. Envelope spectrum of the unfiltered signal from the output shaft sensor

This leads to a conclusion that for cases where the severity of the fault has passed the incipient stage the use of SK, although effectual, is unnecessary. Comparably good results could be obtained by using a simple envelope analysis. Taking this finding into consideration the fault detection algorithm can be simplified, by removing the calculation of the SK for each measurement. With the removal of SK, the selection whether a particular experimental run belongs to the group of fault–free runs or not, can be done based on the amplitudes of spectral components of bearing principle frequencies.

However for the cases of mild faults or for cases of smaller load, the analysis of the envelope spectrum of unfiltered signal turns to be inefficient. This can be seen on the second observed case where the maximal SK value was 4. The envelope spectrum of the unfiltered signal is shown in Figure 15. From the figure we can notice that the signal is dominated by two spectral components located at meshing gears frequencies, unlike the spectrum of the filtered signal shown in Figure 13 which is dominated by the spectral component at f_{BSF}.

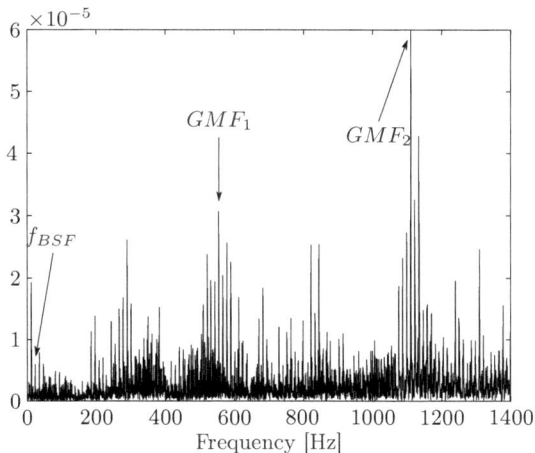

Figure 15. Envelope spectrum of the filtered signal from the input shaft sensor

Therefore we can conclude that for experimental runs conducted with severely damaged bearings the calculation of SK is unnecessary. However for cases of minor damages or cases where the experimental runs were conducted under smaller loads, the calculation of SK and filtering the vibration signals prior to the analysis of the envelope spectrum is an essential step for obtaining the satisfactory results.

5. Conclusion

The problem of fault detection in a case when there is no reference fault–free data, like in the case of the PHM Data Challenge 2009, was resolved using a two step approach: combining spectral kurtosis method and analysis of envelope spectrum. In achieving our goal of bearing fault detection, the gained results of the calculation of SK were twofold. Firstly, the maximum value of SK was used as an indication whether the observed experimental run is likely to represent a run with a faulty element. Secondly, the method selects a frequency band where the impulses generated by the localized faults are most visible. This is very important step, since the amplitude of the localized bearing faults are usually significantly lower then the amplitudes of the vibrations produced by meshing gears, which effectively mask their presence. Thus a simple spectral analysis is ineffective in the process of their detection. So by filtering out all other vibration influences, only spectral components originating from the localized bearing faults remain in the envelope spectrum, which additionally simplifies the fault detection procedure.

Despite all listed results, the general approach based solely on the maximal value of SK has its drawbacks. In cases of multiple faults, there may occur several frequency bands with similarly high values for SK. In such cases all of them should be examined and not just the one with the highest value (Combet & Gelman, 2009).

Regarding the vibration measurements obtained from PHM'09 Data Challenge, the effectiveness of SK can not be fully expressed. This is mainly due to the fact that most of the presented faults had already passed the incipient stage. As it was shown, in such cases the standard envelope analysis of the complete unfiltered signal produces satisfactory results. However, the SK pre–processing approach is effective for the cases of smaller machine loads and mild and incipient faults.

Acknowledgments

The research of the first author was supported by Ad futura Programme of the Slovene Human Resources and Scholarship Fund. We also like to acknowledge the support of the Slovenian Research Agency through Research Programme P2-0001. The authors would like to thank Nader Sawalhi from the School of Mechanical and Manufacturing Engineering, The University of New South Wales, for providing the scripts for Spectral Kurtosis.

References

Antoni, J. (2006). The spectral kurtosis: application to the vibratory surveillance and diagnostics of rotating machines. *Mechanical Systems and Signal Processing, 20*, 308-331.

Antoni, J. (2007). Fast computation of the kurtogram for the detection of transient faults. *Mechanical Systems and Signal Processing, 21*, 108-124.

Antoni, J., & Randall, R. (2006). The spectral kurtosis: a useful tool for characterising non-stationary signals. *Mechanical Systems and Signal Processing, 20*, 282-307.

Bartelmus, W. (2001). Mathematical modelling and computer simulations as an aid to gearbox diagnostics. *Mechanical Systems and Signal Processing, 15*, 855-871.

Benko, U., Petrovčič, J., Musizza, B., & Juričić, Đ. (2008). A System for Automated Final Quality Assessment in the Manufacturing of Vacuum Cleaner Motors. In *Proceedings of the 17th World Congress The International Federation of Automatic Control* (p. 7399-7404).

Combet, F., & Gelman, L. (2009). Optimal filtering of gear signals for early damage detection based on the spectral kurtosis. *Mechanical Systems and Signal Processing, 23*, 652-668.

Dwyer, R. (1983). Detection of non-Gaussian signals by frequency domain Kurtosis estimation. *Acoustics, Speech, and Signal Processing, IEEE International Conference on ICASSP, 8*, 607-610.

Endo, H., Randall, R., & Gosselin, C. (2009). Differential diagnosis of spall vs. cracks in the gear tooth fillet region: Experimental validation. *Mechanical Systems and Signal Processing*, *23*, 636–651.

Ho, D., & Randall, R. B. (2000). Optimisation of bearing diagnostic techniques using simulated and actual bearing fault signals. *Mechanical Systems and Signal Processing*, *14*(5), 763-788.

PHM. (2009). *Prognostics and Health Managment Society 2009 Data Challenge.* http://www.phmsociety.org/competition/09.

Randall, R. B., Antoni, J., & Chobsaard, S. (2001). The relationship between spectral correlation and envelope analysis in the diagnostics of bearing faults and other cyclostationary machine signals. *Mechanical Systems and Signal Processing*, *15*, 945 - 962.

Rubini, R., & Meneghetti, U. (2001). Application of the envelope and wavelet transform and analyses for the diagnosis of incipient faults in ball bearings. *Mechanical Systems and Signal Processing*, *15*(2), 287-302.

Sawalhi, N., & Randall, R. (2008). Simulating gear and bearing interactions in the presence of faults Part I. The combined gear bearing dynamic model and the simulation of localised bearing faults. *Mechanical Systems and Signal Processing*, *22*, 1924-1951.

Sawalhi, N., Randall, R., & Endo, H. (2007). The enhancement of fault detection and diagnosis in rolling element bearings using minimum entropy deconvolution combined with spectral kurtosis. *Mechanical Systems and Signal Processing*, *21*, 2616-2633.

Staszewski, W. J. (1998). Wavelet based compression and feature selection for vibration analysis. *Journal of Sound and Vibration*, *211*, 735 - 760.

Tandon, N., & Choudhury, A. (1999). A review of vibration and acoustic measurement methods for the detection of defects in rolling element bearings. *Tribology International*, *32*, 469-480.

Wang, W. (2001). Early Detection of gear tooth cracking using the resonance demodulation technique. *Mechanical Systems and Signal Processing*, *15*, 887-903.

Information Reconstruction Method for Improved Clustering and Diagnosis of Generic Gearbox Signals

Fangji Wu[1, 2], Jay Lee[2]

[1] State Key Laboratory for Manufacturing Systems Engineering, Research Institute of Diagnostics and Cybernetics,
Xi'an Jiaotong University, Xi'an, Shaanxi, 710049, China
wfjridc@gmail.com

[2] NSF I/UCR Center for Intelligent Maintenance System, University of Cincinnati, OH, 45221, USA
jay.lee@uc.edu

ABSTRACT

Gearbox is a very complex mechanical system that can generate vibrations from its various elements such as gears, shafts, and bearings. Transmission path effect, signal coupling, and noise contamination can further induce difficulties to the development of diagnostic system for a gearbox. This paper introduces a novel information reconstruction approach to clustering and diagnosis of gearbox signals in varying operating conditions. First, vibration signal is transformed from time domain to frequency domain with Fast Fourier Transform (FFT). Then, reconstruction filters are employed to sift the frequency components in FFT spectrum to retain the information of interest. Features are further extracted to calculate the coefficients of the reconstructed energy expression. Then, correlation analysis (CA) and distance measurement (DM) techniques are utilized to cluster signals under diverse shaft speeds and loads. Finally, energy coefficients are used as health indicators for the purpose of fault diagnosis of the rotating elements in the gearbox. The proposed method was used to solve the gearbox problem of the 2009 PHM Conference Data Analysis Competition and won with the best score in both professional and student categories.[*]

1. INTRODUCTION

Gearbox is one of the most widespread and crucial rotating mechanical systems in modern industry. It provides a speed-torque conversion from a higher speed motor to a slower but more forceful output or vice-versa. A gearbox

usually consists of rotating elements such as gears, shafts, and bearings and static elements such as box body and bearing caps. During operation, a gearbox system can suffer the following: gear failures such as wear, scoring, interference, surface fatigue, plastic flow and fracture; bearing failures such as wear, scoring, surface fatigue and brinelling; and shaft failures such as fatigue cracking and overload (Forrester 1996). All these defects can worsen the operating condition and excite excess vibration, and potentially cause major unexpected breakdowns and safety issues. Condition monitoring and fault prognostics of gearbox system have been used for many applications to some degree of success (Peng and Chu 2004, Suh et al. 1999, Wang et al. 2007, Byington et al. 2004). The major challenge is to effectively and accurately identify abnormal patterns early with a sound estimation of the remaining useful life (RUL).

The 2009 PHM Conference Data Analysis Competition is focused on the detection and magnitude estimation of mechanical faults from a generic gearbox using accelerometer data and information about bearing geometry. Participants are scored based on their ability to correctly identify fault type, location, magnitude and damage in the gear system. Data were collected at 30, 35, 40, 45 and 50 Hz shaft speed while being subjected to either high or low loading. Additionally, repeated runs are included in the data, although the run time and load were not sufficient to induce significant fault progression. There are a total of 560 vibration data files to be classified and diagnosed. Details of the Data Analysis Competition are provided on the website http://www.phmsociety.org/competition/09.

This paper introduces a novel information reconstruction approach for clustering and diagnosis of gearbox signals in varying operating conditions.

Fig. 1 is a schematic diagram of the proposed approach. First, vibration signal is transformed from time domain to frequency domain with Fast Fourier Transform (FFT). Second, reconstruction filters are employed to sift the frequency components in FFT spectrum to retain the information of interest and eventually obtain the reconstructed FFT spectrum. Features are further extracted from the modified spectrum to calculate the coefficients of the reconstructed energy expression (energy fitting model). Then, correlation analysis (CA) and distance measurement (DM) techniques are used for clustering signals under diverse shaft speeds and loads. Finally, energy coefficients are used as health indicators for fault diagnosis of the rotating elements in the gearbox. Basically, this approach is a hybrid of data-driven and model-driven schemes. It can be applied as a systematic method for gearbox health assessment system.

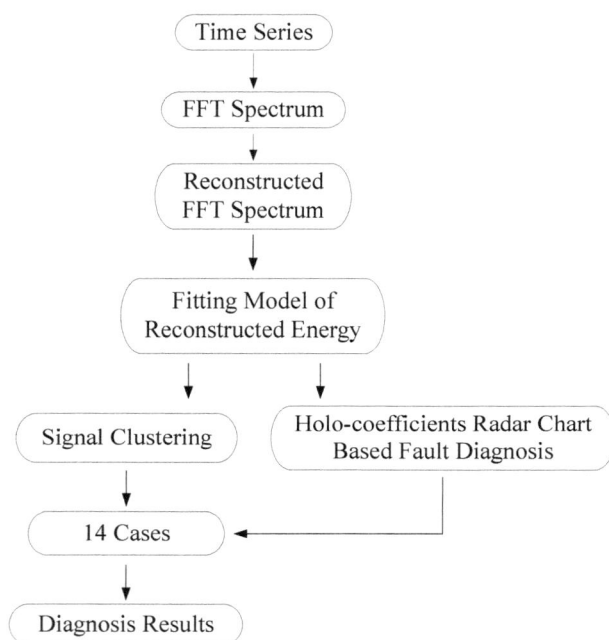

Fig.1. Overview of information reconstruction method

This paper is organized as follows. In Sec. 2, the scheme of reconstructing FFT spectrum is introduced. The feature extraction and reconstructed energy are presented in Sec. 3. Sec. 4 shows the signal clustering process and result of accelerometer data. Sec. 5 introduces holo-coefficients map for gearbox fault diagnosis. The generalization and improvement of the information reconstruction method is discussed in Sec. 6. Finally, conclusions are presented in Sec. 7.

2. RECONSTRUCTED FFT SPECTRUM

To gain further understanding of the gearbox signals, many tools have been developed. These tools consist of time synchronous average (Dempsey 2004) and autoregressive moving average (Wang and Wong 2002)

model for time domain analysis; FFT (Lin et al. 1993), power spectrum (Baydar and Ball 2000), and cepstrum (Badaoui et al. 2001) for frequency domain analysis; short-time Fourier transform (Pinnegar and Mansinha 2003), Wigner-Ville distribution (Baydar and Ball 2000), wavelet transform (Sung et al. 2000), and Hilbert-Huang Transform (Huang et al. 1998) for time-frequency analysis, among others. For the 2009 PHM competition case, vibration data were collected using accelerometers mounted on both the input and output shaft retaining plates. The signal can be described as a complicated measurement with a wide-range energy distribution. However, only some parts of signal are related to specific machine conditions. The main idea of spectrum analysis is to either look at the whole spectrum or look closely at certain frequency components of interest and then extract features from the signal.

To remove or reduce noise and effects from other unexpected sources and further enhance signal components of interest, a reconstruction approach is used to filter and assemble the frequency components to reconstruct signal without loss of information of interest. The scheme of reconstruction method is illustrated in Fig. 2. Each signal is transformed to FFT spectrum. Then, eighteen band-pass filters are applied to select specific frequency bands within the signal. Finally, all the eighteen frequency segments are reassembled together to reconstruct a new signal. The functions of these eighteen band-pass filters are listed in Table 1, which shows the criteria for defining these filters. In this table, frequency components are obtained by calculating corresponding vibration characteristic frequencies of shafts, gears and bearings. Frequency order is the ratio of the characteristic frequency to the shaft rotating frequency.

For shaft, defects such as unbalance and bend will excite harmonic frequency components of shaft rotating frequency. For gear, characteristic frequencies are gear meshing frequency (GMF) and its side band frequencies. GMF is equal to the number of teeth multiplied by the rotational frequency of the gear. It is the periodic signal at the tooth-meshing rate due to deviations from the ideal tooth profile. Side band signals are induced by amplitude modulation effects due to variations in tooth loading; frequency modulation effects due to rotational speed fluctuations and non-uniform tooth spacing; and additive impulses associated with tooth faults. For bearing, a defect on the inner or outer race will cause an impulse each time a rolling element contacts the defect. For an inner race defect this occurs at the inner race ball pass frequency (BPFI), and for an outer race defect this occurs at outer race ball pass frequency (BPFO). A defect on rolling element will cause an impulse each time the defect surface contacts the inner or outer races, which will excite the ball spin frequency (BSF). These characteristic frequencies can be expressed as:

$$
\begin{cases}
BPFI = \dfrac{N}{2}(f_o - f_i)(1 + \dfrac{d}{D}\cos(\alpha)) \\[2mm]
BPFO = \dfrac{N}{2}(f_o - f_i)(1 - \dfrac{d}{D}\cos(\alpha)) \\[2mm]
BSF = (f_o - f_i)(\dfrac{D}{d} - \dfrac{d}{D}\cos^2(\alpha))
\end{cases} \qquad (1)
$$

where N is the number of rolling elements, f_o is the rotational frequency of the outer race, f_i is the rotational frequency of the inner race, d is the diameter of the rolling elements, D is the pitch circle diameter, α is the contact angle.

Table 2 lists the corresponding meaning of these eighteen filters and shows why these filters are defined. The i-X GMF means i-th harmonic frequency of gear meshing frequency. To cite an example, Fig. 3 shows the FFT spectrum of input side signal of File-29 and Fig. 4 shows its reconstructed FFT spectrum.

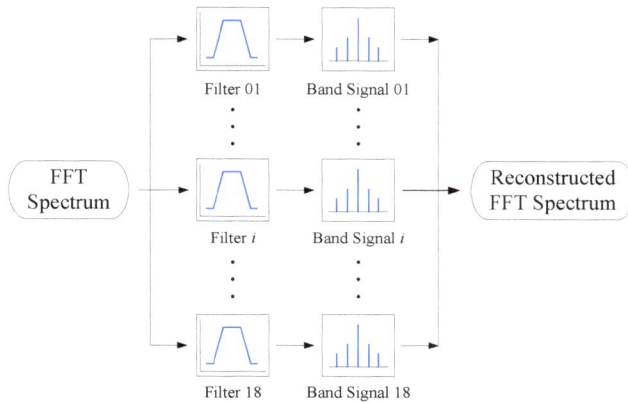

Fig.2. FFT spectrum reconstruction

Table 1. Functions of reconstruction filters

Filter 01	Retaining 01X order component
Filter 02	Retaining 02X order component
Filter 03	Retaining 03X order component
Filter 04	Retaining 04X order component
Filter 05	Retaining 05X order component
Filter 06	Retaining 45X order component
Filter 07	Retaining 06X-10X order component
Filter 08	Retaining 14X-18X order component
Filter 09	Retaining 22X-26X order component
Filter 10	Retaining 30X-34X order component
Filter 11	Retaining 38X-42X order component
Filter 12	Retaining 46X-50X order component
Filter 13	Retaining 54X-58X order component
Filter 14	Retaining 62X-66X order component
Filter 15	Retaining 78X-82X order component
Filter 16	Retaining 94X-98X order component
Filter 17	Retaining 110X-114X order component
Filter 18	Retaining 126X-130X order component

Table 2. Corresponding meaning of filter functions

Filter 01	Characteristic frequency component of input shaft unbalance
Filter 02	Characteristic frequency component of bent input shaft
Filter 03	Characteristic frequency component of outer race defect of input-shaft bearing
Filter 04	Characteristic frequency component of ball defect of input-shaft bearing
Filter 05	Characteristic frequency component of inner race defect of input-shaft bearing
Filter 06	Natural frequency of rotating element or gear ghost frequency component
Filter 07	Output-shaft helical 1X GMF
Filter 08	Input-shaft helical 1X GMF Output-shaft helical 2X GMF Output-shaft spur 1X GMF
Filter 09	Output-shaft helical 3X GMF
Filter 10	Input-shaft helical 2X GMF Output-shaft helical 4X GMF Input-shaft spur 1X GMF Output-shaft spur 2X GMF
Filter 11	Output-shaft helical 5X GMF
Filter 12	Input-shaft helical 3X GMF Output-shaft helical 6X GMF Output-shaft spur 3X GMF
Filter 13	Output-shaft helical 7X GMF
Filter 14	Input-shaft helical 4X GMF Output-shaft helical 8X GMF Input-shaft spur 2X GMF Output-shaft spur 4X GMF
Filter 15	Input-shaft helical 5X GMF Output-shaft spur 5X GMF
Filter 16	Input-shaft helical 6X GMF Input-shaft spur 3X GMF Output-shaft spur 6X GMF
Filter 17	Input-shaft helical 7X GMF Output-shaft spur 7X GMF
Filter 18	Input-shaft helical 8X GMF Input-shaft spur 4X GMF Output-shaft spur 8X GMF

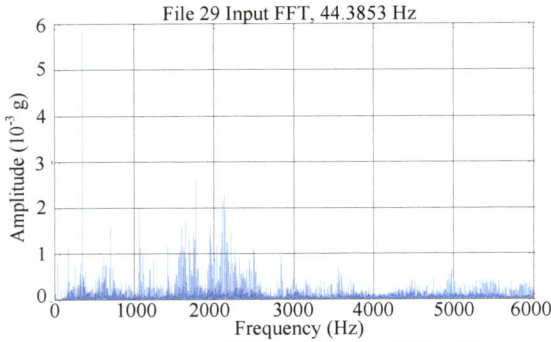

Fig.3. FFT spectrum of File-29

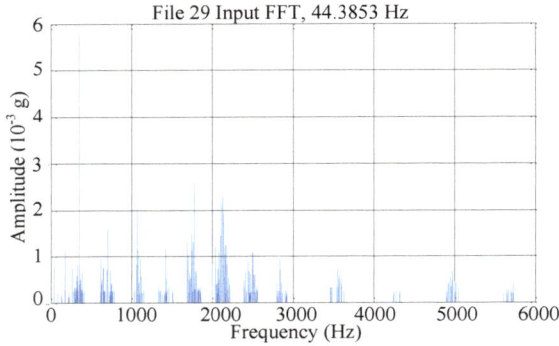

Fig.4. Reconstructed FFT spectrum of File-29

3. FEATURE EXTRACTION AND RECONSTRUCTED ENERGY

Based on the reconstructed FFT spectrum, eighteen features are extracted and they serve as coefficients in the reconstructed energy model. The reconstructed energy can be expressed as:

$$\begin{cases} f_E = f_{EI} + f_{EO} \\ f_{EI} = (\alpha_1 + \cdots + \alpha_6) \times E_{Imax} + (\alpha_7 + \cdots + \alpha_{18}) \times E_{Iall} \\ f_{EO} = (\beta_1 + \cdots + \beta_6) \times E_{Omax} + (\beta_7 + \cdots + \beta_{18}) \times E_{Oall} \end{cases} \quad (2)$$

where f_E is the total energy index of input and output side signals, f_{EI} is the energy index of input side signal, f_{EO} is the energy index of output side signal, E_{Imax} and E_{Omax} are the maximum energy components of input and output side signals, E_{Iall} and E_{Oall} are the full energy values of input and output side signals, α_1 to α_6 are derived by dividing the energy of the first six band signals of input side signal by E_{Imax}, β_1 to β_6 results from dividing energy of first six band signals of output side signal by E_{Omax}, α_7 to α_{18} are computed when energy of last twelve band signals of input side signal is divided by E_{Iall}, and finally, β_7 to β_{18} are determined by dividing the energy of last twelve band signals of output side signal by E_{Oall}.

In the reconstructed energy expression, energy coefficients are selected to have certain classification power. The basic idea is to identify and further classify the data with similar attributes to a specific group. For example, α_1 is supposed

to classify the data either to unbalance group or normal group. Moreover, energy coefficients are also supposed to be comprehensible for user or have physical meaning. This is necessary whenever the classified pattern is to be used for supporting a decision to be made. If the classified pattern is a group without explanation, the user may not trust it. In this paper, knowledge comprehensibility can be achieved by using high-level knowledge representations described in the previous section.

4. SIGNAL CLUSTERING

Given a set of data items, partitioning this set into subsets, such that items with similar characteristics or features are grouped together, is the general idea of signal clustering (Goebel and Gruenwald 1999). A natural way of signal clustering is based on certain similarity measure or distance measure between two signals. In this section, CA and DM on energy coefficients are introduced and evaluated for clustering signals under diverse shaft speeds and loads. Vector of energy coefficients can be constructed as

$$C_E = [\alpha_1, ..., \alpha_{18}, \beta_1, ..., \beta_{18}]^T \quad (3)$$

Then, CA on two signals is defined as

$$CA = (C_{Ei} \cdot C_{Ej}) / (|C_{Ei}| * |C_{Ej}|) \quad (4)$$

where $\cdot$ means dot product, $|\cdot|$ means the largest singular value of a vector. The result of CA ranges between zero and one, with higher CA signifying a higher correlation. DM on two signals is

$$DM = \|C_{Ei} - C_{Ej}\| \quad (5)$$

where $\|\cdot\|$ is the Euclidean distance, with lower DM signifying a higher similarity.

4.1 Determination of Repeated Runs

Using the tachometer signal, rotating speed can be calculated as shown in Fig. 5. There are five distinct groups corresponding to the 5 shaft speeds and each group contains exactly 112 data points. Repeated runs identification was then applied to each speed regime. Consider 50 Hz speed regime, CA for File-157 on these 112 files is illustrated in Fig. 6, while DM for the same scenario is shown in Fig. 7. CA shows that File-183, File-227 and File-498 have the largest correlation value to File-157 and they can be considered as its repeated runs. DM also shows that these three files have the smallest distance value to File-157 and confirms that they are its repeated runs.

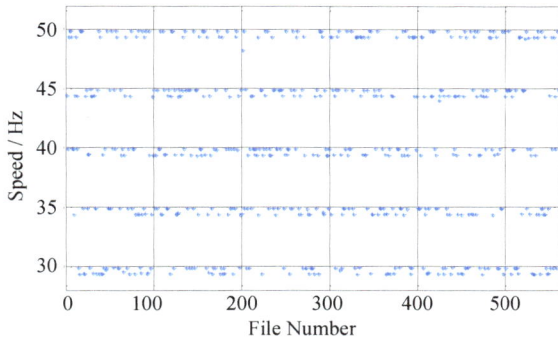

Fig.5. Input shaft speeds

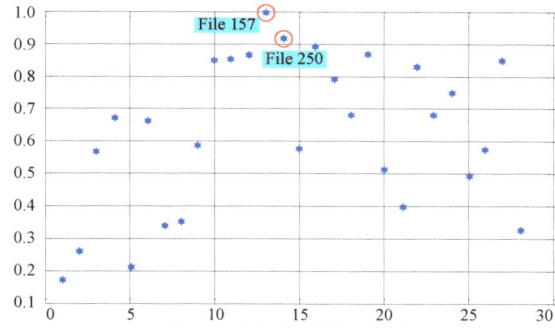

Fig.6. CA for File-157

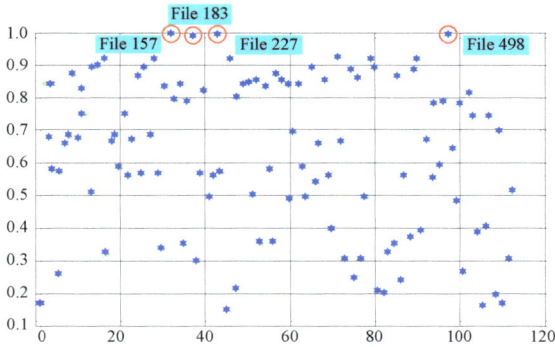

Fig.7. DM for File-157

4.2 Identification of Diverse Loading Runs

After identifying the 4 repeated runs, the 112 files in 50Hz regime are now clustered into 28 groups. CA for File-157 on 28 files from these 28 groups, one file from each group, is illustrated in Fig. 8. DM for File-157 on these 28 files is illustrated in Fig. 9. CA shows that File-250 has the largest correlation value to File-157 and they are from the same pattern. DM shows that File-250 has the smallest distance value to File-157 and they are from the same pattern, one with high load and the other with low load. After identifying the high and low loading runs, 112 files in each speed regime are reduced into 14 groups.

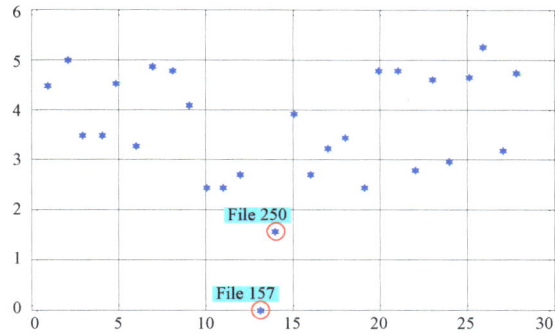

Fig.8. CA for File-157

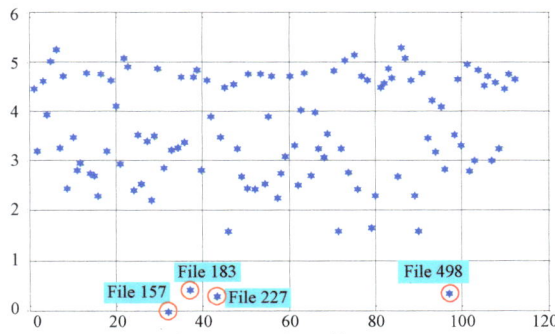

Fig.9. DM for File-157

4.3 Identification of Diverse-Speed Runs

At this point, each speed regime has 14 groups (replications and loading, considered). This section will then describe how the 14 unique patterns are identified across the 5 speed regimes. Consider File 157 (with File-250 as its load pair) in 50Hz regime, its CA and DM with 28 files (one from each of the 28 groups in the same speed regime after identifying replications) in 45Hz regime, are illustrated in Fig. 10 and 11, respectively. Both figures show that File-157 and File-55 (File-62 was its load pair as determined in a previous step) share the same pattern. By doing the same process for the other 3 speed regimes, it was found that File-59, File-69 in 30Hz, File-34, File-88 in 35Hz, File-56, File-213 in 40Hz, File-55, File-62 in 45Hz, and, File-157, File-250 in 50Hz can be clustered as one pattern (Pattern A).

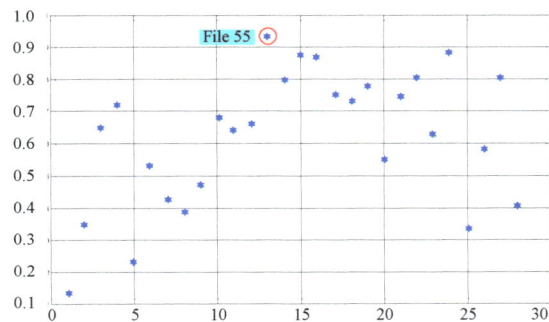

Fig.10. CA for File-157

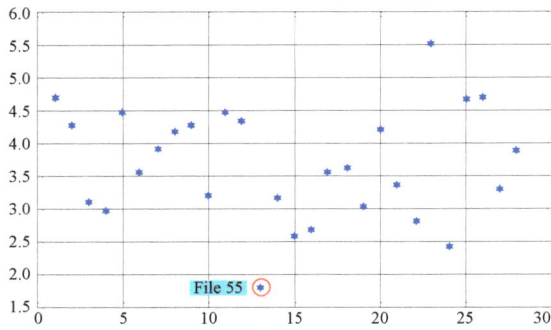

Fig.11. DM for File-157

5. HOLO-COEFFICIENTS MAP/RADAR CHART AND FAULT DIAGNOSIS

The fault diagnosis of rotating elements in the gearbox is performed using energy coefficients as health indicators. A holo-coefficients map comprises of all the energy coefficients. In the map (e.g. Fig. 12 and Fig. 14), the contribution rate of each coefficient can be revealed very clearly along with operating conditions. A more advanced format of holo-coefficients map is holo-coefficients radar chart. The multivariate data in holo-coefficients map are displayed in holo-coefficients radar chart starting from the same point and in different equi-angular spokes, with each spoke representing one of the variables. The data length of a spoke is proportional to the magnitude of the variable for the data point. In the chart (e.g. Fig. 13 and Fig. 15), radial 1 to 18 correspond to α_1 to α_{18}, and radial 19 to 36 correspond to β_1 to β_{18}. The map and chart can be treated as qualitative tools for fault diagnosis. The rules that authors used for qualitative diagnosis are: 1) energy coefficient of a defect should be higher than normal case; the threshold of faulty case depends highly on gearbox set and its dynamic characteristics; usually an energy coefficient larger than 0.4 should trigger a warning, 2) bearing defect may excite lower energy coefficient compared to shaft and gear defect; 3) a high energy coefficient in hard working condition such as high loading and high speed is more reliable for fault detection. Moreover, holo-coefficients map can be updated for quantitative diagnosis. This will be further discussed in next section for generalization of the proposed approach.

Fig. 12 shows the holo-coefficients map of files of Pattern A. Fig. 13 is the transformed radar chart format of Fig. 12. From the figure, input shaft unbalance (radials 1 and 19) and bearing outer defect at input shaft output side (radial 21) are diagnosed. The unbalance excites 1X frequency component as measured from the input side signal and this component is also distinct in output side signal due to the transmission effect of the rigid gearbox housing. The contribution rate of coefficient 3 in 40 Hz is also considerable. However, with the increase in speed, its contribution decreases. Fig. 14 shows the holo-coefficients map of Pattern B (File-60 in 30Hz, File-19 in 35Hz, File-185 in 40Hz, File-36 in 45Hz, and File-258 in 50Hz). Fig. 15 is the transformed radar chart format of Fig. 14. It is

determined that this pattern contains gear error defect at idler shaft 2 location (radials 8 and 26).

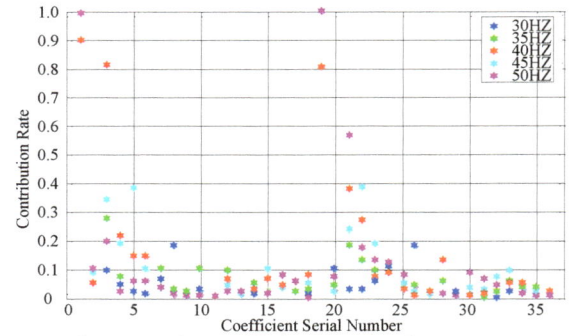

Fig.12. Holo-coefficients map of pattern A

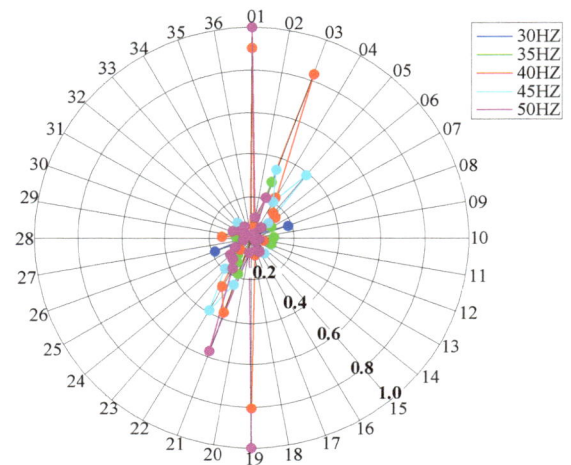

Fig.13. Holo-coefficients radar chart of pattern A

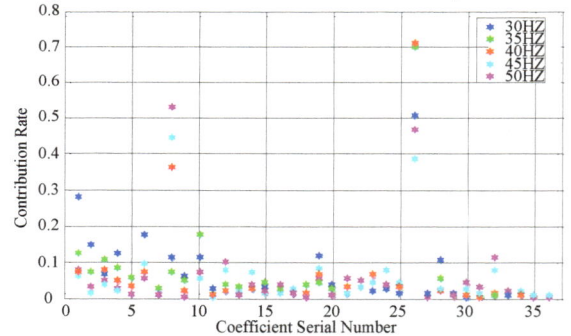

Fig.14. Holo-coefficients map of pattern B

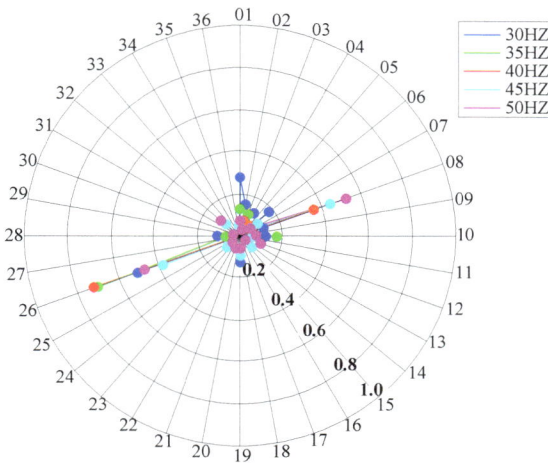

Fig.15. Holo-coefficients radar chart of pattern B

6. GENERALIZATION AND IMPROVEMENT OF INFORMATION RECONSTRUCTION METHOD

The information selection and feature extraction are the crucial steps of the proposed information reconstruction method. The effect of feature selection are (1) to improve classification and diagnosis performance; (2) to visualize the data for model construction; (3) to reduce dimensionality and (4) to remove noise. Improper selection of information of interest and poor extraction of features can lead to under-fitting and over-fitting issues during model creation of the PHM activity of a gearbox system. In developing the energy expression, there is a risk of generating too many energy coefficients which is called over-fitting. Over-fitting will decrease the efficiency and accuracy of the classification since irrelevant attributes can confuse the data mining algorithm. On the contrary, under-fitting means energy coefficients are not enough to support the decision making process.

For over-fitting, it is desirable to have a procedure to prune the ensemble of energy coefficients while keeping the expected classification performance and avoiding the risks in feature selection. The method for selection of energy coefficients that was discussed in this paper relied on expert knowledge which is user-driven and domain-dependent. Had the data files been labeled *a priori,* original files can then be taken as training data, therefore, objective methods, which are data-driven and domain-independent, can be employed to optimize the energy coefficients. The principal component analysis (PCA) can be used to prune the energy coefficients. Because of its ability to discriminate directions with the largest variance in a data set, it is suitable to use PCA for identifying the most representative features. One can first classify data files by pattern; then, apply PCA to feature vectors of data files in each pattern to find the most representative features for the corresponding pattern; finally, assemble retained features from each pattern to obtain the final feature set. Fisher criterion can be another approach for pruning the energy coefficients. Suppose that we have a set of features

in the pattern labeled ω_1 and another set of features in the pattern labeled ω_2. Fisher criterion method actually tries to find the feature set to maximize the distance between two patterns and minimize the deviation within each pattern. A Fisher criterion score can be expressed as:

$$ SF_i = \left| \frac{m(\omega_1)_i - m(\omega_2)_i}{\sigma^2(\omega_1)_i + \sigma^2(\omega_2)_i} \right| \qquad (6) $$

where $m(\omega_1)_i$ and $m(\omega_2)_i$ are the mean value for the i-th feature in ω_1 and ω_2 pattern, $\sigma^2(\omega_1)_i$ and $\sigma^2(\omega_2)_i$ are the standard deviation. By deleting features with small Fisher criterion score, one can exclude irrelevant features from original feature set. Moreover, other advanced feature selection methods such as support vector machine (SVM) and genetic algorithm (GA) based approaches can also be applied (Bradley and Mangasarian 1998, Yang and Honavar 1997).

For under-fitting, more efficient signal processing methods are needed to extract more distinguishable features or more information about the gearbox set is needed to define specified attributes such as natural frequency of gears and bearings. In the current energy expression, the weighting coefficients reflecting the relative importance of energy coefficients are same. If there is evidence proving one energy coefficient is more distinguishable than others, the energy expression can be improved further to have more efficient performance and a more accurate diagnosis. Finally, holo-coefficients radar chart is capable for quantitative diagnosis. However, in order to achieve this goal, there are three sub-tasks need to be considered. First, experiment should be carried out in detail to record the relationship between single energy coefficient and single defect. Second, experiment should be carried out in detail to record the relationship between whole energy coefficients and multi-defects. Third, a model need to be established to represent the relationship between energy coefficients and defects, and then a quantitative reference system and thresholds for quantitative diagnosis can be obtained.

7. CONCLUSION

This paper addressed the information reconstruction method for solving the challenging problem of the 2009 PHM Conference Data Analysis Competition. With this method, raw data can be represented by a reconstructed energy model. Then, based on the energy coefficient of this model, signal clustering can be performed for determination of repeated runs, identification of diverse loading runs, and identification of diverse speed runs. Thus, 560 vibration data files can be classified into 14 patterns. For fault diagnosis of rotating elements in the gearbox, holo-coefficients map and radar chart are used. In the map and chart, the contribution rate of each energy coefficient can be revealed very clearly along with operating conditions. Finally, in order to further apply the information reconstruction method to other gearbox sets besides the one used for PHM competition and to further improve the current approach, four issues are discussed as 1) over-fitting issue, 2) under-fitting issue, 3) weighting coefficient, and 4) quantitative diagnosis. The proposed information reconstruction method can further be applied to the gearbox set working in varying working condition such as helicopter gearbox and wind turbine gearbox for signal clustering and fault diagnosis.

For development of gearbox diagnostic system, extraction of features that are less sensitive or not sensitive to working conditions is critical to accuracy; simulation of the problem-solving process of experts to get diagnosis results with computer is critical to efficiency. In the future, solving problems without interference of experts or performing computer-aided pre-diagnosis before resorting to experts could be expected with the further development of intelligent diagnostic systems.

ACKNOWLEDGMENT

The authors would like to thank Prof. Yudi Shen and other researchers at Research Institute of Diagnostics and Cybernetics, Xi'an Jiaotong University for their kind support during the work of the competition.

REFERENCES

B. D. Forrester. (1996). *Advanced Vibration Analysis Techniques for Fault Detection and Diagnosis in Geared Transmission Systems.* Ph. D. Thesis, Swinburne University of Technology, Melbourne, Australia.

Z. Peng and F. Chu. (2004). Application of the wavelet transform in machine condition monitoring and fault diagnostics: a review with bibliography, *Mechanical Systems and Signal Processing*, vol. 18, pp. 199-221.

J.H. Suh et al. (1999). Machinery fault diagnosis and prognosis: application of advanced signal processing techniques, *CIRP Annals – Manufacturing Technology*, vol. 48, issue 1, pp. 317-320.

J.-Z. Wang, et al. (2007). Gearbox fault diagnosis and prediction based on empirical mode decomposition scheme, *Proceedings of the Sixth International Conference on Machine Learning and Cybernetics*, pp. 1072-1075.

C.S. Byington, et al. (2004). Data-driven neural network methodology to remaining life predictions for aircraft actuator components, *IEEE Aerospace Conference Proceedings*, vol. 6, pp. 3581-3589.

P. J. Dempsey (2004). A Comparison of Vibration and Oil Debris Gear Damage Detection Methods Applied to Pitting Damage, *in Proceedings of COMADEM 2000, 13th International Congress on Condition Monitoring and Diagnostic Engineering Management*, Houston, TX.

W. Wang and A. K. Wong. (2002). Autoregressive Model-Based Gear Fault Diagnosis, *Journal of Vibration and Acoustics*, vol. 124, pp. 172-179.

H. H. Lin, D. P. Townsend, F. B. Oswald (1993). Prediction of Gear Dynamics Using Fast Fourier Transform of Static Transmission Error, *Mechanics Based Design of Structures and Machines: An International Journal*, vol. 21, pp. 237-260.

N. Baydar, A. Ball (2000). Detection of gear deterioration under varying load conditions by using the instantaneous power spectrum, *Mechanical Systems and Signal Processing*, vol. 14, pp. 907-921.

M. E. Badaoui et al. (2001). Use of the moving cepstrum integral to detect and localise tooth spalls in gears, *Mechanical Systems and Signal Processing*, vol. 15, pp. 873-885.

C. R. Pinnegar, L. Mansinha (2003). Time-local spectral analysis for non-stationary time series: The S-Transform for noisy signals, *Fluctuation and Noise Letters*, vol. 3, pp. 357-364.

N. Baydar, A. Ball (2000). A comparative study of acoustic and vibration signals in detection of gear failures using Wigner-Ville distribution, *Mechanical Systems and Signal Processing*, vol. 15, pp. 1091-1107.

C. K. Sung, H. M. Tai and C. W. Chen. (2000). Locating defects of a gear system by the technique of wavelet transform, *Mechanism and Machine Theory*, vol. 35, pp. 1169-1182.

N.E. Huang, et al. (1998). The empirical mode decomposition and the Hilbert spectrum for nonlinear and non-stationary time series analysis, *Proceedings of the Royal Society of London*, Series A 454 pp. 903–995.

M. Goebel, L. Gruenwald (1999). A survey of data mining and knowledge discovery software tools, *ACM SIGKDD Explorations Newsletter*, vol. 1, pp. 20-33.

P. S. Bradley, O. L. Mangasarian. Feature selection via concave minimization and support vector machines, *in Proceedings of 13th International Conference on Machine Learning*, San Francisco, CA.

J. Yang and V. Honavar. Feature subset selection using a genetic algorithm, Feature Extraction, Construction and Selection: A Data Mining Perspective, pp. 117-136, 1998, second printing, 2001.

Fangji Wu is currently a Ph.D. student in State Key Laboratory for Manufacturing Systems Engineering, Research Institute of Diagnostics and Cybernetics, Xi'an Jiaotong University in China, and a visiting scholar in IMS Center, Department of Mechanical Engineering at University of Cincinnati in US. His current research focuses on component-level and system-level CBM and PHM; health management system design and industrial applications; data-driven and model-based methods for fault detection, diagnosis, and prognosis.

Jay Lee is Ohio Eminent Scholar and L.W. Scott Alter Chair Professor at the Univ. of Cincinnati and is founding director of National Science Foundation (NSF) Industry/University Cooperative Research Center (I/UCRC) on Intelligent Maintenance Systems which is a multi-campus NSF Center of Excellence between the Univ. of Cincinnati (lead institution), the Univ. of Michigan, and the Univ. of Missouri-Rolla. His current research focuses on autonomic computing, embedded IT and smart prognostics technologies for industrial and healthcare systems, design of self-maintenance and self-healing, systems, and dominant design tools for product and service innovation. He is also a Fellow of ASME, SME, as well as International Society of Engineering Asset Management (ISEAM).

Comparison of Two Probabilistic Fatigue Damage Assessment Approaches Using Prognostic Performance Metrics

Xuefei Guan[1], Yongming Liu[2], Ratneshwar Jha[1], Abhinav Saxena[3], Jose Celaya[3], Kai Geobel[4]

[1] Department of Mechanical & Aeronautical Engineering, Clarkson University, Potsdam, NY, 13699-5725, USA
guanx@clarkson.edu
rjha@clarkson.edu

[2] Department of Civil & Environmental Engineering, Clarkson University, Potsdam, NY, 13699-5712, USA
yliu@clarkson.edu

[3] SGT, NASA Ames Research Center, Moffett Field, CA, 94035, USA
abhinav.saxena@nasa.gov
jose.r.celaya@nasa.gov

[4] NASA Ames Research Center, Moffett Field, CA, 94035, USA
kai.goebel@nasa.gov

ABSTRACT

In this paper, two probabilistic prognosis updating schemes are compared. One is based on the classical Bayesian approach and the other is based on newly developed maximum relative entropy (MRE) approach. The algorithm performance of the two models is evaluated using a set of recently developed prognostics-based metrics. Various uncertainties from measurements, modeling, and parameter estimations are integrated into the prognosis framework as random input variables for fatigue damage of materials. Measures of response variables are then used to update the statistical distributions of random variables and the prognosis results are updated using posterior distributions. Markov Chain Monte Carlo (MCMC) technique is employed to provide the posterior samples for model updating in the framework. Experimental data are used to demonstrate the operation of the proposed probabilistic prognosis methodology. A set of prognostics-based metrics are employed to quantitatively evaluate the prognosis performance and compare the proposed entropy method with the classical Bayesian updating algorithm. In particular, model accuracy, precision, robustness and convergence are rigorously evaluated in addition to the qualitative visual comparison. Following this, potential development and improvement for the prognostics-based metrics are discussed in detail.

1. INTRODUCTION

Fatigue damage is a critical issue in many structural and non-structural systems, such as aircraft, critical civil structures, and electronic components. The estimation of the reliability and remaining useful life (RUL) is important in condition-based maintenance of a system so that unit replacements can be done in time prior to catastrophic failures. Several physics-based models have been proposed in order to describe the fatigue process and predict the damage propagation; among these, Paris-type crack growth models (Paris & Erdogan, 1963; Forman et al., 1967; Walker, 1970) are most commonly used (Bourdin, Francfort, & Marigo, 2008). However, experimental data indicate that fatigue crack propagation is not a smooth, stable and well ordered process (Virkler, Hillberry, & Goel, 1979), thus a deterministic model is not capable of quantifying the crack growth subject to various uncertainties associated with the fatigue damage. Uncertainties arising from a number of sources, such as measurement errors, model prediction residuals, and non-optimal parameter estimation, affect the quality of life predictions. These uncertainties need to be carefully included and managed in the prognosis process for risk management and decision-making.

In order to model the stochastic process of fatigue propagation and gain knowledge about a target system via monitoring system responses, probabilistic updating methods based on Bayes theorem have been used to evaluate the probability density functions (PDF) of input parameters using response measurements. For example, see (Madsen & Sorensen, 1990;

Zhang & Mahadevan, 2000). Entropy methods, such as Maximum Entropy (MaxEnt) methods (Jaynes, 1957, 1979; Skilling, 1988) and relative entropy methods (Van Campenhout & Cover, 1981; Haussler, 1997), are alternative approaches for probability assignment and updating and have been used in many applications such as statistical mechanics (Caticha & Preuss, 2004; Tseng & Caticha, 2008), quantum physics (Hiai & Petz, 1991; Vedral, 2002), and fatigue prognosis (Guan, Jha, & Liu, 2009). This paper has two objectives; the first is to develop a general prognosis approach based on maximum relative entropy (MRE) principles for probabilistic fatigue damage prognosis and compare it to the classical Bayesian approach, and the other is to explore prognosis metrics to evaluate prognosis performance quantitatively. One of the advantages of the proposed MRE approach is that the resulting confidence bounds are narrower compared to the classical Bayesian method, which is beneficial for decision making in a health management setting. The rest of the paper is organized as follows. In section 2, we review the classical Bayesian approach and formulate a general MRE updating and prognosis framework. To the best knowledge of the authors, this is the first attempt to apply the MRE method as a general methodology in fatigue damage problems. Section 3 presents two application examples and methodology validation. Section 4 discusses algorithmic performance metrics and extends the two examples of Section 3 in this context. Following that are discussions and conclusion.

2. PROBABILISTIC MODEL UPDATING

In this section, both the classical Bayesian probability updating and a general MRE prognosis framework for fatigue damage problems are introduced. To evaluate the posterior probability distribution, Markov Chain Monte Carlo (MCMC) simulation is then introduced and employed in this framework to approximate the target distribution. For a generic inference problem with an uncertain parameter vector $\theta \in \Theta$, the posterior PDF of θ is inferred on the basis of three pieces of information: the prior knowledge about θ (the prior PDF of θ), the observation of a response event/variable $x \in X$, and the known relationship between x and θ (the likelihood function based on physical/mathematical models). The search space for desired posterior PDF of θ is $X \times \Theta$. Both Bayesian and MRE are capable of performing the search for an optimized posterior. However, these two approaches are based on different mechanisms. This is discussed in details in the following paragraphs.

2.1. Classical Bayesian Model Updating

Bayes' theorem provides a model for inductive inference or the learning process. A Bayesian posterior PDF is a measure of known information about parameters with uncertainty. Bayes' theorem is a means for combining the observation re-

garding the related parameters through the likelihood function (Gregory, 2005). Let $p(\theta)$ denote the prior PDF of θ. According to the Bayes' theorem, the posterior PDF of a variable θ that reflects the fact that we observed x' is

$$p(\theta|x') \propto p(\theta)p(x'|\theta) \qquad (1)$$

The Bayesian formulation of a posterior is straightforward and has an enormous variety of applications. Detailed derivation and demonstration can be found in the referred article and is not repeated here. One issue with the classical Bayesian approach is that only response observations can be used for updating. Other types of information, such as the expected value of a parameter and statistical moments, cannot be directly incorporated into the classical Bayesian framework. For example, coupon level experiment testing and failure analysis can reflect statistical features of batch productions. The statistical information can further help to improve the individual prognosis performance. In order to include this type of information in the probabilistic prognosis and model updating, an entropy-based probabilistic inference framework has been developed. Details are discussed below.

2.2. MRE Approach for Model Updating

The relative information entropy, also referred to as Kullback-Leibler divergence (Kullback & Leibler, 1951), of two PDFs $p_1(\theta)$ and $p_2(\theta)$ is defined as,

$$I(p_1 : p_2) = - \int_\Theta p_1(\theta) \ln \frac{p_1(\theta)}{p_2(\theta)} d\theta, \qquad (2)$$

where θ is the parameter vector and Θ is the associated vector space. The axioms of maximum entropy indicate that the form of Eq. (2) is the unique entropy representation for inductive inference (Skilling, 1988).

The three axioms are:

1. Locality. Local information has local effects.

2. Coordinate invariance. The ranking of the two probability densities should not depend on the system coordinates. This indicates that the coordinates carry no information.

3. Consistency for independent subsystem. For a system composed of subsystems that are independent; it should not make a difference whether the inference treats them separately or jointly.

Using the similar notation above, let $p(x, \theta)$ be a prior joint PDF and $q(x, \theta)$ be the posterior joint PDF. According to the entropy axioms, the selected joint posterior is the one that maximizes the relative entropy $I(q : p)$ in Eq. (3), subject to all available constraints, such as statistical moments and measures of a response variable.

$$I(q:p) = -\int_{X \times \Theta} q(x,\theta) \ln \frac{q(x,\theta)}{p(x,\theta)} dx d\theta. \quad (3)$$

In Eq. (3), $p(x,\theta) = p(x)p(x|\theta)$ contains all prior information, $p(x|\theta)$ is the conditional PDF or likelihood function and $p(\theta)$ is the prior PDF of θ. The same relationship applies to $q(x,\theta)$. When new information is available in the form of a constraint, the updating procedure will search in the space of $X \times \Theta$ for a posterior which maximizes $I(q:p)$. Measurements of the response variable x can be used to perform the updating, which is performed in a similar way as the classical Bayesian updating. The benefit of MRE updating is that it can incorporate other information for inference, which cannot be included in the classical Bayesian updating. For example, the expected value of a function of θ from experiments or the empirical judgment on the mean value of θ. This flexibility of applicable information can pose more constraints on a posterior thus yield a more accurate result given that those constraints are justified. If a new observation x' is obtained, the posteriors that reflect the fact x is now known to be x' is a constraint such that

$$c_1 : q(x) = \int_{\Theta} q(x,\theta)d\theta = \delta(x - x'). \quad (4)$$

Other information in the form of moment constraints, such as the expected value of some function $g(\theta)$, can be formulated as

$$c_2 : \int_{X \times \Theta} q(x,\theta)g(\theta)dx d\theta = \langle g(\theta) \rangle. \quad (5)$$

The normalization constraint is

$$c_3 : \int_{X \times \Theta} q(x,\theta)dx d\theta = 1. \quad (6)$$

Maximizing Eq. (3) using the method of Lagrange multipliers, subject to constraints Eqs. (4-6) and the posterior PDF of θ is obtained as

$$q(\theta) \propto p(\theta)p(x'|\theta)\exp[\beta g(\theta)]. \quad (7)$$

The detailed derivation of Eq. (7) and the computation of the Lagrange multiplier β can be found in (Guan et al., 2009). The right side of Eq. (7) consists of three terms. $p(\theta)$ is the parameter prior, $p(x'|\theta)$ is the likelihood, and $\exp[\beta g(\theta)]$ is the exponential term introduced by moment constraints. Eq. (7) is similar to Bayesian posterior except for the additional exponential term. This equation further indicates that, if no moment constraint is available, i.e., β is zero, MRE updating will be identical to Bayesian updating. In other words, Bayesian updating is a special case of MRE updating. Similar to that of a Bayesian updating problem, the likelihood function is usually constructed using the physics-based model depending on different realistic applications.

2.3. Fatigue Mechanism Model and Likelihood Function Construction

In this section, a general procedure of constructing the likelihood equation is presented. Let d be a response variable measure of our target system and y be the prediction value of a prediction model M. If the model is sufficiently accurate to describe the system output, the observed value is equal to model prediction value, i.e. $y = d$. However, noise and errors usually exist for both modeling and measurements. Incorporating a modeling uncertainty term e and a measurement noise term ϵ into consideration and assuming both errors are additive to obtain

$$d = M(\theta) + e + \epsilon, \quad (8)$$

where $M(\theta)$ is the deterministic model prediction and θ is the associated model parameter variable. Without the evidence that e and ϵ are correlated to each other, the two terms are assumed to be two independent zero-mean normal variables and can be collected as a new normal variable $\tau = (e + \epsilon) \sim \text{Norm}(0, \sigma_\tau)$, the likelihood function for multiple observations can be constructed as

$$p(d_1, \ldots, d_n | \theta) = \frac{1}{(\sqrt{2\pi}\sigma_\tau)^n} \exp\left[-\frac{1}{2} \sum_{i=1}^{n} \left(\frac{d_i - M_i(\theta)}{\sigma_\tau}\right)^2\right]. \quad (9)$$

Substituting Eq. (9) in Eq. (7), the MRE posterior of θ is obtained as

$$p(\theta | d_1, \ldots, d_n) \propto p(\theta)\exp\left[-\frac{1}{2} \sum_{i=1}^{n} \left(\frac{d_i - M_i(\theta)}{\sigma_\tau}\right)^2 + \beta g(\theta)\right]. \quad (10)$$

For fatigue damage model $M(\theta)$, various deterministic models have been proposed to describe the fatigue crack accumulation, among which Paris type of models are commonly used in cycle based fatigue crack growth calculation. In this study, Paris model (Paris & Erdogan, 1963) is employed for illustration purposes. In a realistic situation, other model might be adopted accordingly. Let a be the crack length, N be the number of cycles, the Paris' model reads,

$$\frac{da}{dN} = c(\Delta K)^m = c[\Delta \sigma \sqrt{\pi a} F(a)]^m, \quad (11)$$

where c and m are model parameters, $\Delta \sigma$ is the stress variation during one cyclic load, ΔK is the variation of stress intensity in one cyclic load, and $F(a)$ is the geometry correction factor. The crack size can be calculated by solving Eq. (11) given the parameter c and m and the applied number of loading cycles N. Early studies have show that $\ln c$ follows a normal distribution and m follows truncated normal distribution (Kotulski, 1998). Given this information, the posterior of the joint distribution of $(\ln c, m)$ can be expressed as,

$$p(\ln c, m) \propto$$

$$\exp\left[-\frac{1}{2}\left(\frac{\ln c - \mu_{\ln c}}{\sigma_{\ln c}}\right)^2 + \beta_{\ln c}g_{\ln c}(\ln c)\right] \times$$

$$\exp\left[-\frac{1}{2}\left(\frac{m - \mu_m}{\sigma_m}\right)^2 + \beta_m g_m(m)\right] \times$$

$$\exp\left[-\frac{1}{2}\left(\frac{d_i - M_i(\ln c, m)}{\sigma_\tau}\right)^2\right].$$

$$(12)$$

Setting $\beta_{\ln c}$ and β_m to zero in Eq. (12) gives the Bayesian formulation of the same problem. The PDF of one parameter can be obtained by integrating over the rest of the parameters. But for a large dimension parameter space, more general and computationally efficient methods, such as sampling techniques, might be applied.

2.4. MCMC Simulation Method

Direct evaluation of the PDF in Eq. (12) is difficult because of the multi-dimensional integration needed for normalization. In order to circumvent the direct evaluation of Eq. (12), Markov Chain Monte Carlo sampling technique is used in this study. MCMC was first introduced by (Metropolis et al., 1953) as a method to simulate a discrete-time homogeneous Markov chain. The merit of MCMC is that it overcomes the normalization of Eq. (12) and ensures that the state of the chain converges to the target distribution after a large number of steps from an arbitrary initial start. The widely used random walk algorithm, Metropolis-Hastings algorithm (Hastings, 1970), is summarized here.

The transition between two successive samples x_t and x_{t+1} is defined by Eq. (13).

$$x_{t+1} = \begin{cases} \tilde{x} \sim \pi(X|x_t) & \text{with probability } \alpha(x_t, \tilde{x}) \\ x_t & \text{else} \end{cases}$$

$$(13)$$

$\pi(X|x_t)$ is the transition distribution, and $\alpha(x_t, \tilde{x}) = \min(1, r)$ is the acceptance probability. The Metropolis ratio r is defined as,

$$r = \frac{p(\tilde{x})}{p(x_t)}\frac{\pi(x_t|\tilde{x})}{\pi(\tilde{x}|x_t)},$$

$$(14)$$

where $p(\cdot)$ is the target distribution. In our case, $p(\cdot)$ is Eq. (12). For a symmetric transition distribution $\pi(\cdot)$, such as a normal distribution, the property of $\pi(x_t|\tilde{x}) = \pi(\tilde{x}|x_t)$ reduces Eq. (14) to $r = p(\tilde{x})/p(x_t)$. In this study, 100,000 samples of $(\ln c, m)$ are generated with a 5% burn-in period using a normal transition distribution. In addition, the moment information of these samples is then integrated into the proposed MRE updating procedure.

3. APPLICATION EXAMPLES

Two fatigue crack growth experimental datasets are used to demonstrate the proposed MRE updating procedure and show the benefits of this approach.

3.1. Virkler's 2024-T3 Aluminum Alloy Experimental Data

An extensive fatigue crack growth data under constant loading for Al 2024-T3 plate specimens with center through cracks was collected in (Virkler et al., 1979). The dataset consists of 68 fatigue crack growth trajectories and each trajectory contains 164 measurement points. All specimens have the same geometry, i.e., an initial crack size $a_i = 9\text{mm}$, length $L = 558.8\text{mm}$, width $w = 152.4\text{mm}$ and thickness $t = 2.54\text{mm}$. The loading information is $\Delta\sigma = 48.28\text{MPa}$ and stress ratio $R = 0.2$. The geometry correction factor for these specimens is $F(a) = 1/\sqrt{\cos(\pi a/w)}$. (Kotulski, 1998) reported the statistical information of the parameters in Paris' model, namely, mean values $\mu_{\ln c} = -26.155$ and $\mu_m = 2.874$ with standard deviations $\sigma_{\ln c} = 0.968$ and $\sigma_m = 0.164$, respectively. Assuming the total error term is $\tau = 0.1\text{mm}$ and substituting the statistics information into Eq. (12) with $g_{\ln c} = \ln c$ and $g_m(m) = m$, the updating procedure can be performed when observation data become available.

One crack growth trajectory in Virkler's dataset was selected arbitrarily for fatigue crack length prediction updating from (Ostergaard & Hillberty, 1983). Five data points in the early stage of the crack propagation are randomly chosen to represent the measured ground truth values of crack length obtained from health monitoring system or nondestructive inspection. These data points are listed in Table 1.

Number	Crack size (mm)	Cycles
1	9.733	21269
2	10.527	42734
3	11.256	56392
4	12.171	73161
5	15.055	110487

Table 1. Data used for updating (Virkler's dataset)

Predictions from MRE updating and Bayesian updating procedures are shown in Figure 1. To keep the figure clear, the median prediction (expected value) is omitted. As can be seen, MRE updating gives a narrower prognosis confidence interval as compared to classical Bayesian updating. It further justifies that the additional moment constraints imposed on the posterior yield a more compact results.

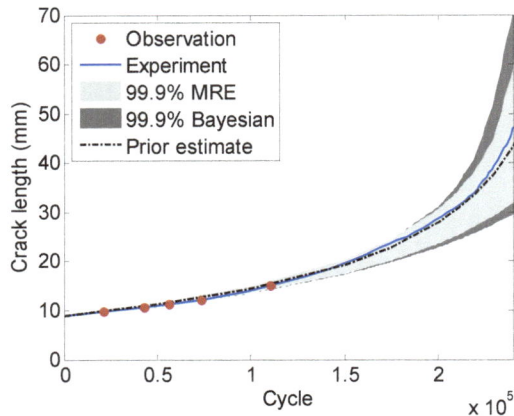

Figure 1. MRE and Bayesian prognosis (Virkler's dataset)

Number	Crack size (mm)	Cycles
1	11.361	4875
2	11.928	8475
3	12.325	11550
4	13.856	17775
5	14.877	21375

Table 2. Data used for updating (McMaster's dataset)

3.2. McMaster's 2024-T351 aluminum alloy experimental data

In (McMaster & Smith, 1999), a large set of 2024-T351 aluminum alloy experimental data under constant and variable loading conditions were reported. The experimental data of center-cracked specimens with length $L = 250$mm, width $w = 100$mm and thickness $t = 6$mm under constant loading $\Delta\sigma = 65.7$MPa and stress ratio $R = 0.1$ are used. Priors of the parameters are obtained by $\ln(da/dN) \sim \ln(\Delta K)$ regression using the experimental data. Five data points as shown in Table 2 are chosen arbitrarily to represent sensor measurements from health monitoring system. The prior PDFs are artificially set as $\mu_{\ln c} = -26.5$ and $\mu_m = 2.9$, which is not sufficiently accurate enough to match the experimental records as seen in Figure 2. Predictions of crack growth trajectories are also shown in Figure 2, where interval predictions obtained by MRE updating are narrower than that by Bayesian updating. One interesting observation is that the difference between MRE and Bayesian interval predictions in Figure 2 is larger than that in Figure 1. One possible explanation is that the prior PDF settings in the two datasets have different level of uncertainties. The prior PDFs in Figure 1 are sufficiently accurate. We can observe this because the prior point estimate in black dash line (computed using the mean value reported by Kotulski) is very close to the experiment records in solid blue line. For the McMaster's dataset, the prior PDFs for the Paris' equation parameters are artificial set. The prior estimate is far from the experiment records. The affect of prior PDFs settings is further discussed in Section 5. In the two examples, MRE updating shows the advantages over Bayesian updating by visual observation. This is more likely due to the additional statistical moment constraints of MCMC samples added to posteriors. To quantify the performance, prognosis metrics need to be considered to provide a rigorous comparison between MRE updating and Bayesian updating as given below.

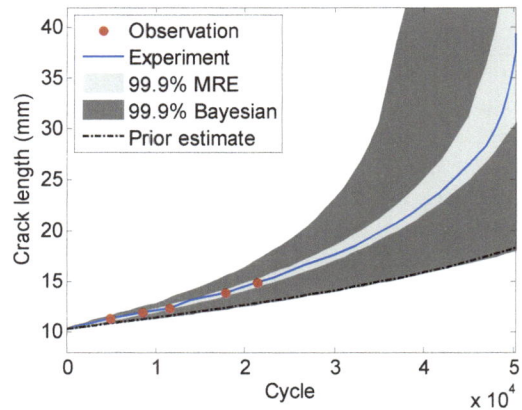

Figure 2. MRE and Bayesian prognosis (McMaster's dataset)

4. METRIC-BASED PERFORMANCE EVALUATION

Various metrics are available to quantify the performance of prognosis algorithms (Saxena, Celaya, Balaban, et al., 2008). In this section, classical error based statistical measures and several prognosis metrics are applied to quantify the prediction performance of application examples in the previous section.

4.1. Statistical Metrics

Metrics, such as mean squared error (MSE), mean absolute percentage error (MAPE), average bias, sample standard deviation (STD), and their variations are widely used in medicine and finance fields where large datasets are available for statistical data analysis (Saxena, Celaya, Balaban, et al., 2008). The results for those classical metrics shown in Table 3 and Table 4 (rows 1-4) are computed using the prediction residuals (the difference between actual RUL and predicted RUL) obtained after the fifth updating. The proposed MRE approach shows its advantages over Bayesian method in all cases.

4.2. Prognosis Metrics

The statistical metrics mentioned above are general purpose metrics and were not specifically designed for prognosis. In (Saxena, Celaya, Saha, Saha, & Goebel, 2008) authors proposed several metrics, such as Prognostic Horizon (PH), Alpha-Lambda ($\alpha - \lambda$) Performance, Relative Accuracy (RA), Cumulative Relative Accuracy (CRA), and Convergence; that were designed specifically for prognosis to incorporate the prediction distributions and the structure of the prognostics process. These metrics help assess how well prediction estimates improve over time as more measurement data become available. For readers' reference, we present a brief definition of these metrics here.

1. Prognostic Horizon is defined as the length of time before end-of-life (EoL) when an algorithm starts predicting within specified accuracy limits. These limits are specified as $\pm\alpha\%$ of the true EoL.

2. $\alpha - \lambda$ Accuracy determines whether predictions from an algorithm are within $\pm\alpha\%$ accuracy of the true RUL at a given time instant, specified by the parameter λ. For instance a $\lambda = 0.5$ would specify midway between the first time a prediction is made and EoL.

3. Relative Accuracy quantifies the percent accuracy with respect to actual RUL at a given time (specified by λ). It's an accuracy measure normalized by RUL, signifying that predictions closer to EoL should be more accurate and precise.

4. Cumulative Relative Accuracy is a weighted average of RAs computed at different time instances. Weights can be assigned to the predictions based on how critical they become as EoL approaches, and hence the accuracy of the predictions.

5. Convergence quantifies the rate at which any performance metric of interest improves to reach its desired value as time passes by.

For more description, implementation details and application examples on these metrics; the reader may referred to (Saxena, Celaya, Saha, et al., 2008). In general, these metrics were designed to capture the time varying aspects of prognostics. As more data become available prognostic estimates get revised. It is, therefore, important to track how well an algorithm performs as time passes by as opposed to evaluating performance at one specific time instant only. Further, these metrics also incorporate the notion of increased criticality as EoL approaches, which imply that a successful prognosis algorithm should improve as the system approaches its EoL. In this paper we compare the two approaches based on Bayesian and MRE updating. In addition to evaluating performance based on prognosis metrics, we also include some classical statistical metrics. For this purpose, in our approach we include an additional updating point from the end of time series to establish EoL and compute the RUL curves. Results obtained from this evaluation exercise are presented next.

Performance Results for Virkler's Dataset

The visual results for PH and $\alpha - \lambda$ accuracy are shown in Figure 3. Numerical values of those metrics are listed in Ta-

ble 3. For computing CRA (see Table 3), the starting point is cycle zero because the specimens have initial cracks. We evaluated RA at 20, 40, 60, and 80% of EoL and did not use weighting factors. This assumes that relative accuracy is equally weighted at all time instants. Though, this may not always be preferable, a simplistic evaluation was carried out to observe the natural behavior of the algorithm itself. Figure 3 compares the prediction horizon for the two algorithms with 10% error bound around EoL value. Using the strict definition for PH as laid out in (Saxena, Celaya, Saha, Saha, & Goebel, 2009), we observed that MRE yields a larger PH. The plot of PH performance in Figure 3 shows that 90% MRE interval prediction enters the 90% accuracy zone at the fifth updating, while Bayesian prediction enters the zone at the sixth updating showing that MRE is slightly better than Bayesian. It is worth mentioning that there is no specific reason to choose , which is very conservative and strict. Typically 50% corresponds to evaluating mean value being inside the alpha bounds. It depends on specific reliability requirement and actual application constraints to pick up a proper value. In general, it indicates that, for engineering practice, the proposed MRE can give an informative prediction at an earlier stage of the whole lifecycle. The statistical metrics, MAPE, Average Bias, STD, MSE, are computed after the fifth updating. The prognosis metrics of PH, RA, and CRA are computed using the 90% interval predictions of RUL at each updating points. For the convergence metric, the median prediction is used here. Looking at Table 3 one can see that on Virkler's dataset MRE performs better than Bayesian approach under all performance measures. One must note that although classical metrics conclude the same as the new prognostics metrics, they do not take into account the time varying nature of the prognostics and hence may not always be useful in practice.

Metrics	MRE	Bayesian
MAPE	8.66	10.93
Average Bias(cycle)	10956.27	14051.92
STD(cycle)	7628.77	9115.78
MSE(cycle^2)	178.23×10^6	280.5×10^6
PH(cycle)	132016	83583
RA$_{\lambda=0.4}$	0.92	0.89
CRA	0.89	0.87
Convergence	74365.72	77349.24

Table 3. Comparison of metrics between MRE and Bayesian approaches (Virkler's dataset, statistical metrics (rows 1-4) are computed after fifth updating)

Performance Results for McMaster's Dataset

A similar analysis for the McMaster's dataset is performed. The visual results for PH and $\alpha - \lambda$ accuracy metrics com-

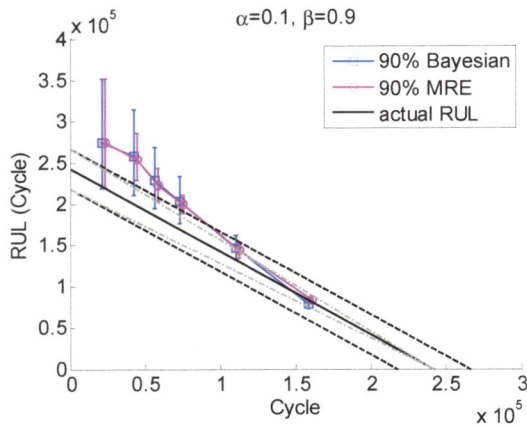

Figure 3. Performance comparison for PH and $\alpha-\lambda$ accuracy at $\alpha = 0.1$ (10% error bound) on Virkler's dataset

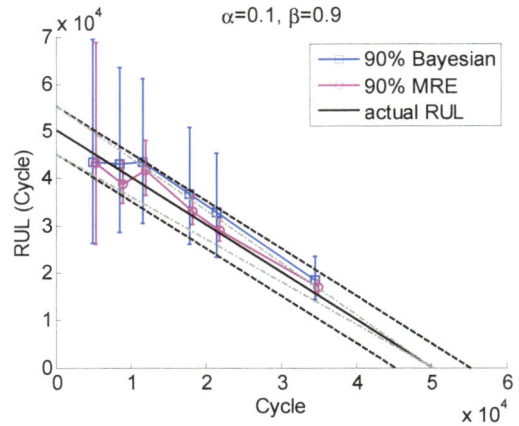

Figure 4. Performance comparison for PH and $\alpha-\lambda$ accuracy at $\alpha = 0.1$ (10% error bound) on McMaster's dataset

paring Bayesian and MRE updating are shown in Figure 4. The rest of the metrics are included in Table 4. The general conclusion about the superior performance of the MRE procedure from Virkler's dataset is further strengthened. The MRE's superior performance over Bayesian approach is attributed to the ability to incorporate additional knowledge about the system using additional constraints. For this dataset, these metrics clearly distinguish the two approaches and show better outcomes from the MRE method. For example, the PH and $\alpha - \lambda$ performance metrics shown in Figure 4 present clear visual comparisons, e.g., the prognosis bounds obtained by MRE enters the cone area at the fourth updating which is earlier than that of Bayesian.

Metrics	MRE	Bayesian
MAPE	4.06	22.53
Average Bias(cycle)	418.76	4561.93
STD(cycle)	1413.53	6888.38
MSE(cycle^2)	2.17×10^6	68.26×10^6
PH(cycle)	32475	N/A
RA$_{\lambda=0.4}$	0.99	0.86
CRA	0.95	0.87
Convergence	13757.94	22175.16

Table 4. Comparisons of metrics between MRE and Bayesian approaches (McMaster's dataset, statistical metrics (rows 1-4) are computed after fifth updating)

5. DISCUSSION

As observed in the previous section, there are a few aspects where these metrics can be further enhanced to improve performance evaluation. The significant difference between the PHs for the two algorithms may also be an artifact of the frequency at which these algorithms make a prediction. We also

observed that in a probabilistic prognosis updating scheme, the selection of priors may produce different prognosis results and affect the performance. Consequently, different updating methods may exhibit different robustness with inappropriate priors. Next, we discuss some of these issues as they relate to prognosis metrics.

5.1. Convergence Metric

The convergence metric computes a value to quantify how fast prognostic estimates improve and converge towards the ground truth. A metric like convergence is meaningful only if an algorithm improves with time and passes various criteria defined by other prognostic metrics. For example, the convergence in terms of RUL relative error (RE) defined in Eq. (15), which is the difference between an actual response measure (R) and the inferred value (R_0) divided by the actual response measure. The result of Virkler's dataset shows a monotonic decreasing trend after the second update (Figure 5). Both MRE and Bayesian methods show diverging trends for McMaster's dataset (Figure 6). The results (converging and diverging trends) suggests that a metric like convergence will not make complete sense if the algorithms do not show improvements with time and hence additional fine tuning of the algorithms is required. The length of the dash line (Figure 5 and Figure 6) between the coordinate origin and the centric point of the area covered by the RE curves serves as a quantitative value of convergence metric. The details of that can be found in (Saxena, Celaya, Saha, et al., 2008). It is worth mentioning that different applications may require different measures instead of RE and the choice of measures depends on which aspect of the algorithmic convergence we would like to investigate.

$$RE := \left| \frac{R_0 - R}{R} \right|. \qquad (15)$$

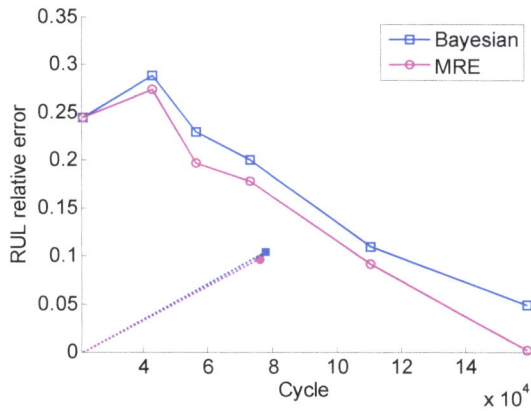

Figure 5. Comparison of convergence performance on Virkler's dataset

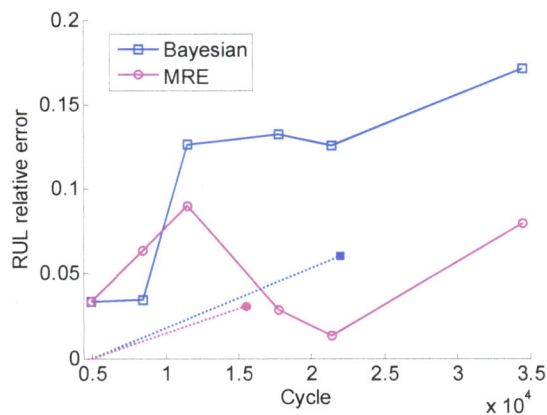

Figure 6. Comparison of convergence performance on McMaster's dataset

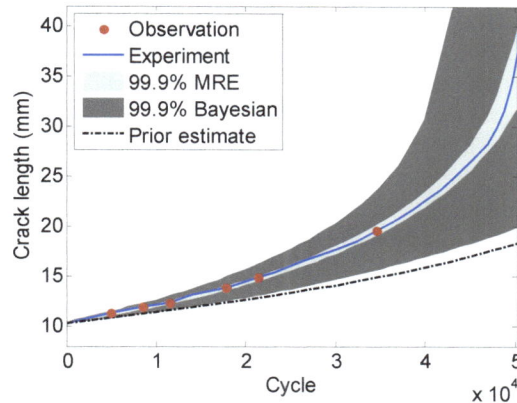

Figure 7. MRE and Bayesian prognosis with an inaccurate prior (McMaster's dataset)

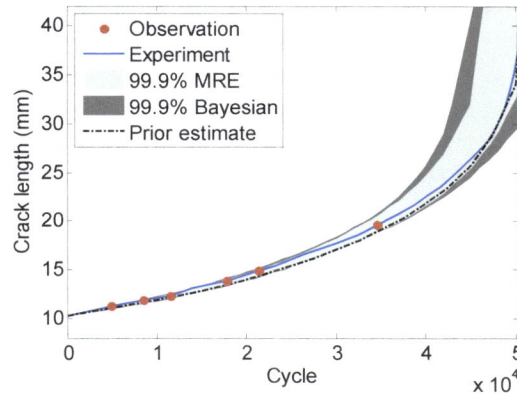

Figure 8. MRE and Bayesian prognosis with an accurate prior (McMaster's dataset)

5.2. Robustness metric

From the above examples, it is shown that the selection of a prior PDF is critical for a meaningful prognosis using probabilistic updating schemes such as Bayesian and MRE. An inaccurate prior may render a poor prediction of RUL. For example, when the prior prediction (shown in Figure 7) is very different from the actual crack growth trajectory, the Bayesian predictions lead to inaccurate estimates with very wide confidence bounds. The MRE updating approach performs well while using the same inaccurate prior distributions. On the other hand, starting with a relatively accurate prior prediction, both MRE and Bayesian give similar predictions as shown in Figure 8. It is valuable to define a robustness metric that can quantify the sensitivity of different algorithms with respect to the algorithm parameters, such as prior distribution, initial conditions, and training data size.

A preliminary study on the robustness metric is shown below. The basic idea is to quantify the change of prognosis confidence bounds due to the changing of algorithm parameter values. The range of investigated parameter is first defined based on specific application requirements (e.g., 10% variation around the mean value) or based on the underlining physics requirement (e.g., parameter should be non-negative). In this paper, we used a parameter η to specify the range of interested parameter (i.e., the parameter is in the range of mean $\pm\eta$). For a robust algorithm, the change of algorithm parameters will not affect the prognosis confidence bounds much. In view of this, the area in a confidence bound vs. parameter variation plot is a good indication of algorithm robustness (shaded area in Figures 9 and 10). In order to perform the metric comparison across different parameter spaces, a normalization process is proposed. A reference area is defined by specifying an allowable prediction error level (e.g., $\pm20\%$ in the current investigation). This allowable level is expressed using parameter δ. The reference area can be calculated as $4\eta\delta$ and is shown as the area by the dashed lines in Figures 9 and 10. Mathematically, the robustness metric R_b can be defined as

$$R_b := \frac{\int_{x_{\text{mean}}-\eta}^{x_{\text{mean}}+\eta} f(x)\,\mathrm{d}x}{4\eta\delta}, \qquad (16)$$

where x is the investigated algorithm parameter and $f(x)$ is the confidence bound variation function with respect to x. The physical meaning of Eq. (16) is the shaded area normalized by the dashed line area in Figures 9 and 10. The performance of the two updating algorithms is investigated using the above mentioned robustness metric for Virkler's dataset first. In this case, $\eta = 0.2$ and $\delta = 0.2$ are used to investigate the parameter m in the crack growth model (Eq. (12)). The mean value of m is 2.874. All predictions are made after six updatings and the 99% confidence bounds are shown in Figure 9. The robustness metric (Eq. (16)) of the Bayesian approach is 2.6 while that for the MRE approach is 0.7. The similar investigation if performed for McMaster's dataset with the mean value of m equaling to 2.9. The robustness metric of the Bayesian and MRE approach are 3.0 and 0.4, respectively. The metric configuration and the visual comparison for McMaster's dataset are shown in Figure 10.

From the above results we can see that, under this specific parameter configuration, MRE exhibits more robust against the variation of m in prior PDFs. In fatigue damage problems, the model parameters are usually tuned using extensive experiments on standard specimens. The realistic systems are usually different from specimens in geometric dimensions, loading profiles, and usage environment. Extensive experiments on the actual engineering systems are sometimes prohibitive due to the time and cost constraints. Therefore, it may be valuable in a practical perspective since most of the time an accurate prior is difficult to obtain with a limited data source. One issue with this robustness metric is that it does not reflect how the performance changes with time. More complicated metrics based on this idea maybe developed by

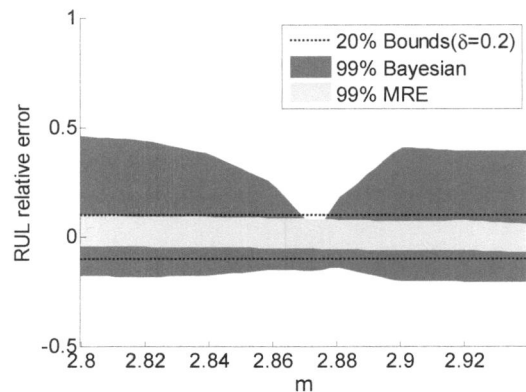

Figure 9. Comparison of robustness metric after six updatings with varying values of m in prior PDF (Eq. 12) for parameter m (Virkler's dataset)

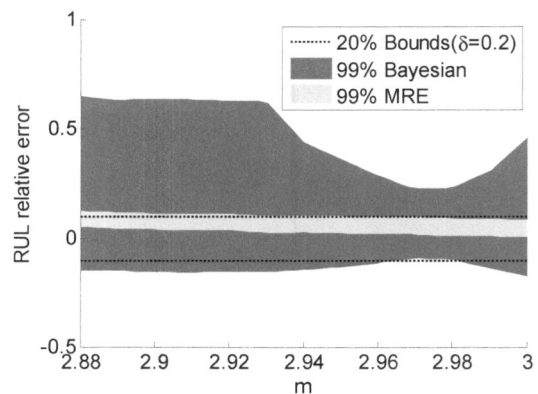

Figure 10. Comparison of robustness metric after six updatings with varying values of m in prior PDF (Eq. 12) for parameter m (McMaster's dataset)

adding another dimension to record the performance variation with time. Since Bayesian updating algorithms are associated with many factors, such as the total number of updating points, the training data size, noise levels, etc., further studies are needed to establish such concepts regarding the algorithmic robustness.

To make further comparison between different Bayesian updating and prognosis approaches, more data points and even the whole dataset can be used as observation data to see with enough measures of response whether MRE and Bayesian give similar prognosis results and show convergence. Though in practice it is more desirable to get an early stage accurate prognosis, it is necessary to explore the characteristics of different updating algorithms using experimental data as we showed in previous sections.

6. CONCLUSION

A general framework for probabilistic prognosis using maximum entropy approach, MRE, is proposed in this paper to include all available information and uncertainties for RUL prediction. Prognosis metrics are used for model comparison and performance evaluation. Several conclusions can be drawn based on the results in the current investigation:

The proposed MRE updating approach results in more accurate and precise prediction compared with the classical Bayesian method.

The classical Bayesian method is a special case of the proposed MRE approach and MRE approach is more flexible to include additional information for inference, which cannot be handled by the classical Bayesian method. The prognosis metrics can be successfully used for algorithm comparison and can give quantitative values in model (algorithm) performance evaluation.

A robustness metric measuring the updating algorithmic sensitivity to prior uncertainty is proposed and applied to both Bayesian and MRE updating approaches. The application examples show that MRE exhibits more robustness against the uncertainty introduced by parameter distribution priors in the sense of prognosis performance.

It is important to realize when to apply these metrics to arrive at meaningful interpretations. For instance, use of the convergence metric makes sense only when the algorithm predictions converge (get better) with time.

ACKNOWLEDGMENT

The research reported in this paper was supported in part by the NASA ARMD/AvSP IVHM project under NRA NNX09AY54A. The support is gratefully acknowledged.

NOMENCLATURE

$I(\cdot)$	Relative information entropy
$p(\cdot)$	Probability distribution
$M(\cdot)$	Fatigue crack growth model
$F(\cdot)$	Geometry correction factor
a	Crack length
N	Number of loading cycles
d	Crack length measurements

REFERENCES

Bourdin, B., Francfort, G., & Marigo, J. (2008). The variational approach to fracture. *Journal of Elasticity*, *91*(1), 5–148.

Caticha, A., & Preuss, R. (2004). Maximum entropy and bayesian data analysis: Entropic prior distributions. *Physical Review E*, *70*(4), 046127.

Forman, R., Kearney, V., & Engle, R. (1967). Numerical analysis of crack propagation in cyclic-loaded structures. *Journal of Basic Engineering*, *89*(3), 459–464.

Gregory, P. (2005). *Bayesian logical data analysis for the physical sciences: a comparative approach with mathematica support*. Cambridge Univ Pr.

Guan, X., Jha, R., & Liu, Y. (2009). Probabilistic fatigue damage prognosis using maximum entropy approach. *Journal of Intelligent Manufacturing*, 1-9. (DOI: 10.1007/s10845-009-0341-3)

Hastings, W. (1970). Monte Carlo sampling methods using Markov chains and their applications. *Biometrika*, *57*(1), 97.

Haussler, D. (1997). A general minimax result for relative entropy. *Information Theory, IEEE Transactions on*, *43*(4), 1276–1280.

Hiai, F., & Petz, D. (1991). The proper formula for relative entropy and its asymptotics in quantum probability. *Communications in mathematical physics*, *143*(1), 99–114.

Jaynes, E. (1957). Information theory and statistical mechanics. ii. *Physical review*, *108*(2), 171.

Jaynes, E. (1979). Where do we stand on maximum entropy. *The maximum entropy formalism*, 15–118.

Kotulski, Z. (1998). On efficiency of identification of a stochastic crack propagation model based on virkler experimental data. *Archives of Mechanics*, *50*, 829–848.

Kullback, S., & Leibler, R. (1951). On information and sufficiency. *The Annals of Mathematical Statistics*, 79–86.

Madsen, H., & Sorensen, J. (1990). Probability-based optimization of fatigue design, inspection and maintenance. *Integrity of offshore structures—4*, 421.

McMaster, F., & Smith, D. (1999). Effect of load excursions and specimen thickness on crack closure measurements. *Advances in Fatigue Crack-closure Measurements and Analysis*, 246–264.

Metropolis, N., Rosenbluth, A., Rosenbluth, M., Teller, A., Teller, E., et al. (1953). Equation of state calculations by fast computing machines. *The journal of chemical physics*, *21*(6), 1087.

Ostergaard, D., & Hillberty, B. (1983). Characterization of the variability in fatigue crack propagation data. *Probabilistic Frature Mechanics and Fatigue Methods: Applications for Structural Design and Maintenance*, 97.

Paris, P., & Erdogan, F. (1963). A critical analysis of crack propagation laws. *Journal of Basic Engineering*, *85*(4), 528–534.

Saxena, A., Celaya, J., Balaban, E., Goebel, K., Saha, B., Saha, S., & Schwabacher, M. (2008). Metrics for evaluating performance of prognostic techniques. In *International conference on prognostics and health management (phm08)*.

Saxena, A., Celaya, J., Saha, B., Saha, S., & Goebel, K.

Model Adaptation for Prognostics in a Particle Filtering Framework

Bhaskar Saha[1] and Kai Goebel[2]

[1]Mission Critical Technologies, Inc. (NASA ARC), 2041 Rosecrans Avenue, Suite 220, El Segundo, CA 90245
bhaskar.saha@nasa.gov

[2]NASA Ames Research Center, Moffett Field, CA 95134, USA
kai.goebel@nasa.gov

ABSTRACT

One of the key motivating factors for using particle filters for prognostics is the ability to include model parameters as part of the state vector to be estimated. This performs model adaptation in conjunction with state tracking, and thus, produces a tuned model that can used for long term predictions. This feature of particle filters works in most part due to the fact that they are not subject to the "curse of dimensionality", i.e. the exponential growth of computational complexity with state dimension. However, in practice, this property holds for "well-designed" particle filters only as dimensionality increases. This paper explores the notion of wellness of design in the context of predicting remaining useful life for individual discharge cycles of Li-ion batteries. Prognostic metrics are used to analyze the tradeoff between different model designs and prediction performance. Results demonstrate how sensitivity analysis may be used to arrive at a well-designed prognostic model that can take advantage of the model adaptation properties of a particle filter.

1. INTRODUCTION

The field of system health management (SHM) is undergoing a paradigm shift from the reliability driven maintenance strategies that relied on metrics like mean-time-to-failure (MTTF), to more proactive condition-based maintenance (CBM) strategies that estimate the remaining useful life (RUL) specific to the system under consideration. This results in more efficient performance, longer system life, as well as reduction in costs from unscheduled maintenance due to unforeseen failures. The applicability of this methodology that was once pioneered by the aerospace and the defense industry now ranges far and wide from green buildings to electric cars to consumer electronics.

The trigger for this evolution has been the concept of *prognostics* and the need to integrate it into the operations and maintenance decisioning process. The definition of what constitutes prognostics is still an open discussion in the SHM community, but for the purposes of this paper, we will define it to be *the process by which the evolution of a system variable or vector indicating its health is tracked over time under current and proposed future usage, until its value no longer falls within the limits set forth by the system specifications.* This somewhat broadens the definition set forth by Saxena *et al.* (2008), where prognostics is triggered by a diagnostic routine, and the detected failure precursor is tracked through time until a predefined end-of-life (EOL) threshold is reached. Other applications may include predicting nominal wear or intermediate cycle-life as discussed in the case of rechargeable batteries by Saha & Goebel (2009).

Prognostic approaches can be broadly classified into two categories: *data-driven* and *model-based*. Data-driven techniques mainly exploit evolution trends of the tracked variable observed from training or archived data under similar operational conditions. Although, they circumvent the need for domain expertise and model development both of which cost time and money, they lead to the problem of data availability and integrity. In most cases, little data is collected from engineered systems in use. This may not be true for aerospace applications, but even when there is data, very little of it is actually collected under faulty conditions. Accelerated aging tests are even more rare since most systems are either too costly to run to failure, or take too long to do so. Additionally, there are problems with sensor bias and drift, and in some cases, outright failure.

This motivates the development of model-based techniques where domain expertise may be brought to bear. However, most high fidelity models are too computationally intractable to be run in an online environment that can be integrated with the decisioning process. Consequently, there is a need for a model-based prognostic framework that can track the nonlinear dynamics of system health while using a lower-order system representation. The *Particle Filter* (PF) introduced by Gordon *et al.* (1993) is an elegant solution to this need. PFs are a novel class of nonlinear filtering methods that combine *Bayesian learning* techniques with *importance sampling* to provide good state tracking performance. Additionally, model parameters can be included as a part of the state vector to be tracked, thus performing model adaptation in conjunction with state estimation. The model, thus tuned during the tracking phase, can then be propagated subject to expected future use to give long-term prognosis.

2. BACKGROUND

Nonlinear filtering has been an active topic of research for the last several decades in the statistical and engineering community (Jazwinski, 1970). The core problem is to sequentially estimate the state of a dynamic system $\mathbf{x}_k \in \mathbb{R}^{n_x}$, where $\mathbb{R}$ is the set of real numbers and n_x is the dimension of the state vector, using a time-sequence of noisy measurements $\mathbf{z}_k \in \mathbb{R}^{n_z}$, where n_z is the dimension of the measurement vector (Ristic *et al.*, 2004). The time index $k \in \mathbb{N}$, where $\mathbb{N}$ is the set of natural numbers, is assigned to the continuous-time instant t_k. Thus the state evolution model and the measurement equation may be expressed as:

$$\mathbf{x}_k = \mathbf{f}_{k-1}(\mathbf{x}_{k-1}, \omega_{k-1}) \qquad (1)$$

$$\mathbf{z}_k = \mathbf{h}_k(\mathbf{x}_k, \nu_k) \qquad (2)$$

where, $\mathbf{f}$ and $\mathbf{h}$ are known nonlinear functions, and ω and ν represent process and measurement noise sequences, possibly non-Gaussian, whose statistics are known. It is desired to obtain the filtered estimates of $\mathbf{x}_k$ from all available measurements $\mathbf{Z}_k \equiv \{\mathbf{z}_i, i = 1,\dots,k\}$

up to t_k, which, from a Bayesian perspective, amounts to constructing the *posterior* pdf (probability density function) $p(\mathbf{x}_k|\mathbf{Z}_k)$. Once the *initial* density $p(\mathbf{x}_0) \equiv p(\mathbf{x}_0|\mathbf{Z}_0)$ is determined, the pdf may be obtained recursively using the prediction and update steps shown in Eqs. (1) and (2).

Let us say that at time t_{k-1} we have the pdf $p(\mathbf{x}_{k-1}|\mathbf{Z}_{k-1})$. In the prediction step the system model in Eq. (1) is used to obtain the *prior* pdf at time t_k via the Chapman-Kolmogorov equation:

$$p(\mathbf{x}_k|\mathbf{Z}_{k-1}) = \int p(\mathbf{x}_k|\mathbf{x}_{k-1}, \mathbf{Z}_{k-1}) p(\mathbf{x}_{k-1}|\mathbf{Z}_{k-1}) d\mathbf{x}_{k-1} \,. \qquad (3)$$

Assuming a first-order Markov process, $p(\mathbf{x}_k|\mathbf{x}_{k-1}, \mathbf{Z}_{k-1}) = p(\mathbf{x}_k|\mathbf{x}_{k-1})$, which may be determined from Eq. (1) and the known statistics of ω_{k-1}. Equation (3) thus reduces to:

$$p(\mathbf{x}_k|\mathbf{Z}_{k-1}) = \int p(\mathbf{x}_k|\mathbf{x}_{k-1}) p(\mathbf{x}_{k-1}|\mathbf{Z}_{k-1}) d\mathbf{x}_{k-1} \,. \qquad (4)$$

At time t_k when the measurement $\mathbf{z}_k$ is received, the prior pdf is updated using Bayes' rule as follows:

$$
\begin{aligned}
p(\mathbf{x}_k|\mathbf{Z}_k) &= p(\mathbf{x}_{k-1}|\mathbf{z}_k, \mathbf{Z}_{k-1}) \\
&= \frac{p(\mathbf{z}_k|\mathbf{x}_k, \mathbf{Z}_{k-1}) p(\mathbf{x}_k|\mathbf{Z}_{k-1})}{p(\mathbf{z}_k|\mathbf{Z}_{k-1})} \\
&= \frac{p(\mathbf{z}_k|\mathbf{x}_k) p(\mathbf{x}_k|\mathbf{Z}_{k-1})}{p(\mathbf{z}_k|\mathbf{Z}_{k-1})} \,.
\end{aligned} \qquad (5)
$$

The last step of Eq. (5) assumes that the measurements are independent of each other such that $\mathbf{z}_k$ only depends upon $\mathbf{x}_k$. The normalizing constant in the denominator can be represented in terms of the *likelihood* function $p(\mathbf{z}_k|\mathbf{x}_k)$, defined by Eq. (2) and the known statistics of ν, as follows:

$$p(\mathbf{z}_k|\mathbf{Z}_{k-1}) = \int p(\mathbf{z}_k|\mathbf{x}_k) p(\mathbf{x}_k|\mathbf{Z}_{k-1}) d\mathbf{x}_k \,. \qquad (6)$$

Substituting Eq. (6) into Eq. (5), we can express the *posterior* pdf obtained after the update step as:

$$p(\mathbf{x}_k|\mathbf{Z}_k) = \frac{p(\mathbf{z}_k|\mathbf{x}_k) p(\mathbf{x}_k|\mathbf{Z}_{k-1})}{\int p(\mathbf{z}_k|\mathbf{x}_k) p(\mathbf{x}_k|\mathbf{Z}_{k-1}) d\mathbf{x}_k} \,. \qquad (7)$$

The recurrence relations in Eqs. (4) and (7) form the basis for computing the optimal Bayesian estimate. However, these integrals are rarely ever analytical in nature, thus leading to the need for sub-optimal filters like particle filters. PFs evaluate these integrals by performing Monte Carlo (MC) integration, which is the basis for all sequential Monte Carlo (SMC) estimation methods. Noting the fact that $\int p(\mathbf{x}_{k-1}|\mathbf{Z}_{k-1})d\mathbf{x}_{k-1} = \int p(\mathbf{x}_k|\mathbf{Z}_{k-1})d\mathbf{x}_k = 1$, both the integrals in Eqs. (4) and (7) can expressed in the form of:

$$I = \int \Phi(\mathbf{x})\pi(\mathbf{x})d\mathbf{x} \qquad (8)$$

where, $\pi(\mathbf{x})$ is of the form $p(\mathbf{x}_{k-l}|\mathbf{Z}_{k-1})$, $l = 0$ or 1, satisfying the pdf properties $\pi(\mathbf{x}) \geq 0$ and $\int \pi(\mathbf{x})d\mathbf{x} = 1$. $\Phi(\mathbf{x})$ may be derived from Eqs. (1) and (2) for Eqs. (4) and (7) respectively. The MC estimate of this integral can expressed as the mean of $N \gg 1$ samples $\{\mathbf{x}^i; i = 1,\dots,N\}$:

$$I_N = \frac{1}{N}\sum_{i=1}^{N}\Phi(\mathbf{x}^i). \qquad (9)$$

Assuming independent samples, I_N is an unbiased estimate and, according to the *law of large numbers*, will converge to I. Given the fact that in our case $\Phi(\mathbf{x})$ is a pdf constrained within the values of 0 and 1, its variance $\sigma^2 = \int (\Phi(\mathbf{x})-I)^2\pi(\mathbf{x})d\mathbf{x}$ is also finite. This means that applying the *central limit theorem* the estimation error can be said to converge as:

$$\lim_{N\to\infty}\sqrt{N}(I_N-I) \sim \mathcal{N}(0,\sigma^2) \qquad (10)$$

where $\mathcal{N}(0,\sigma^2)$ denotes a normal distribution with zero mean and variance σ^2. The MC estimate error, $e = I_N - I$, is of the order of $O(N^{1/2})$, which means that the rate of convergence is dependent on the number of particles N, but not the dimension of the state, n_x (Ristic *et al.*, 2004). This leads to the notion that PFs are not subject to the *curse of dimensionality* like other nonlinear filters.

The phrase "curse of dimensionality" was coined by Richard Bellman (1957) more than half a century ago to denote the exponential increase in computational complexity in nonlinear filters as a function of the state dimension n_x. Daum (2005) in his tutorial on nonlinear filters discusses this aspect of particle filters. He states that "It has been asserted that PFs avoid the curse of dimensionality, but this is generally incorrect. Well designed PFs with good proposal densities sometimes avoid the curse of dimensionality, but not otherwise." Figure 1 and Figure 2, reprinted from (Daum, 2005), show the comparison between the median

dimensionless error for good and poor proposal densities respectively evaluated over a chosen nonlinear filtering problem with "vaguely Gaussian" conditional densities (Daum & Huang, 2003).

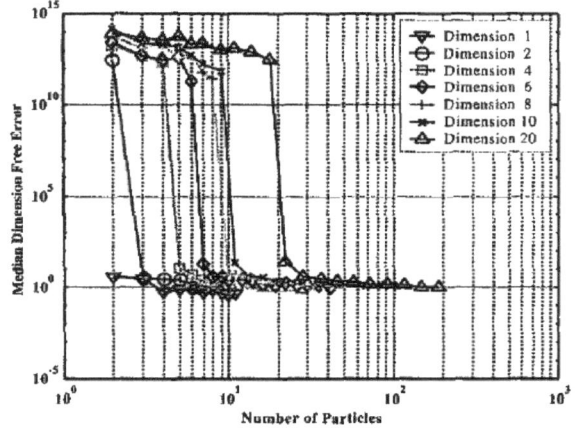

Figure 1. Dimension free error vs. number of particles for PF with good proposal density (Daum, 2005).

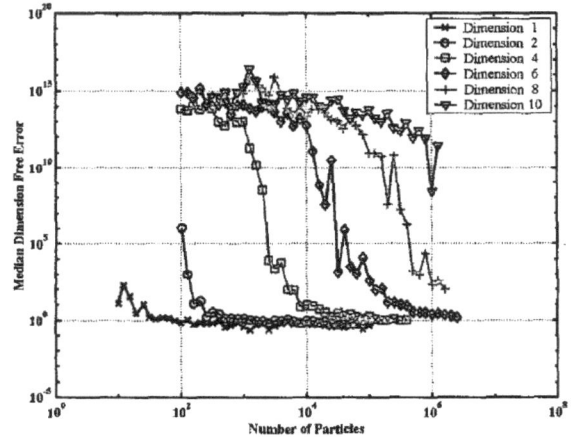

Figure 2. Dimension free error vs. number of particles for PF with poor proposal density (Daum, 2005).

It can be seen from the figures that for a state vector of dimension 8 i.e., $n_x = 8$, the PF with the poor proposal density achieves the same error level with about 10^6 particles that a PF with good proposal density achieves with 10 particles. This discrepancy gets exponentially higher as the dimensionality of the state vector increases linearly, clearly showing that the PF performance does not always escape the curse of dimensionality. Further discussion on this topic can be found in (Daum & Huang, 2003).

The theoretical basis behind the particle filter escaping the curse of dimensionality is that the proposal density considered, given by the samples $\{\mathbf{x}^i; \; i = 1,...,N\}$, come from the regions of the state space that are important for the pdf integration results in Eqs. (4) and (7). However, it is usually not possible to sample effectively from the posterior distribution $\pi(\mathbf{x})$ being multivariate, non-parametric and, in most cases, unknown beyond a proportionality constant (Ristic *et al.*, 2004). In the case of the prognostic problem, even though the system health vector to be tracked may not be high dimensional, the incorporation of model parameters into the state vector, in order to track the non-stationarity of the system model, adds extra dimensions (Saha & Goebel, 2009). Thus, model adaptation that facilitates good prognosis necessitates a good choice of proposal density.

3. THE PROGNOSTICS FRAMEWORK

Before we investigate the issues with model adaptation, let us take a step back and look at how prognostics is performed in the PF framework. The framework has been described before (Saha *et al.*, 2009), however, some basic elements are reproduced below in order to set the context. Particle methods assume that the state equations can be modeled as a first order Markov process with additive noise and conditionally independent outputs. Under these assumptions Eqs. (1) and (2) become:

$$\mathbf{x}_k = \mathbf{f}_{k-1}(\mathbf{x}_{k-1}) + \omega_{k-1} \quad (11)$$

$$\mathbf{z}_k = \mathbf{h}_k(\mathbf{x}_k) + \mathbf{v}_k . \quad (12)$$

As mentioned in (Daum, 2005) there are several flavors of PFs. Analyzing all is not within the scope of this paper. Here we shall focus on *Sampling Importance Resampling* (SIR), which is a very commonly used particle filtering algorithm that approximates the posterior filtering distribution denoted as $p(\mathbf{x}_k|\mathbf{Z}_k)$ by a set of N weighted particles $\{\langle x_p^i, w_p^i \rangle; \; i = 1,...,N\}$ sampled from a distribution $q(\mathbf{x})$ that is "similar" to $\pi(\mathbf{x})$, i.e., $\pi(\mathbf{x}) > 0 \Rightarrow q(\mathbf{x}) > 0$ for all $\mathbf{x} \in \mathbb{R}^{n_z}$. The *importance weights* w_k^i are normalized in the following way:

$$w_k^i = \frac{\pi(\mathbf{x}_k^i)/q(\mathbf{x}_k^i)}{\sum\limits_{j=1}^{N} \pi(\mathbf{x}_k^j)/q(\mathbf{x}_k^j)} \quad (13)$$

such that $\Sigma_i w_k^i = 1$, and the posterior distribution can be approximated as:

$$p(\mathbf{x}_k|\mathbf{Z}_k) \approx \sum_{i=1}^{N} w_k^i \delta(\mathbf{x}_k - \mathbf{x}_k^i). \quad (14)$$

Using the model in Eq. (11) the prediction step from Eq. (4) becomes:

$$p(\mathbf{x}_k|\mathbf{Z}_{k-1}) \approx \sum_{i=1}^{N} w_{k-1}^i \mathbf{f}_{k-1}(\mathbf{x}_{k-1}^i). \quad (15)$$

The weights are updated according to the relation:

$$\overline{w}_k^i = w_{k-1}^i \frac{p(\mathbf{z}_k|\mathbf{x}_k^i) p(\mathbf{x}_k^i|\mathbf{x}_{k-1}^i)}{q(\mathbf{x}_k^i|\mathbf{x}_{k-1}^i, \mathbf{z}_k)}, \quad (16)$$

$$w_k^i = \frac{\overline{w}_k^i}{\sum\limits_{j=1}^{N} \overline{w}_k^i} . \quad (17)$$

Resampling is used to avoid the problem of degeneracy of the PF algorithm, i.e., avoiding the situation that all but a few of the importance weights are close to zero. If the weights degenerate, we not only have a very poor representation of the system state, but we also spend valuable computing resources on unimportant calculations. More details on this are provided in (Saha *et al.*, 2009). The basic logical flowchart is shown in Figure 3.

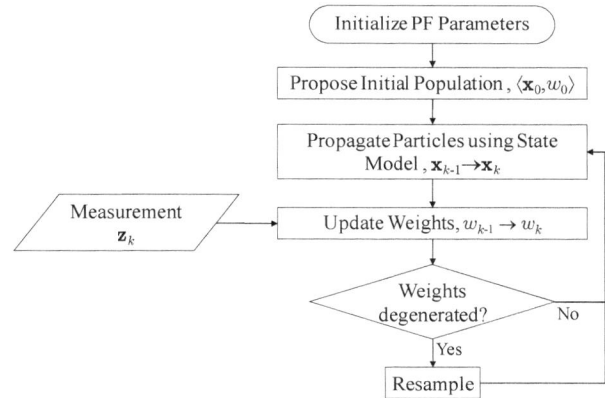

Figure 3. Particle filtering flowchart.

During prognosis this tracking routine is run until a long-term prediction is required, say at time t_p, at which point Eq. (11) will be used to propagate the posterior pdf given by $\{\langle x_p^i, w_p^i \rangle; \; i = 1,...,N\}$ until $\mathbf{x}^i$ fails to meet the system specifications at time t_{EOL}^i. The RUL pdf, i.e., the distribution $p(t_{EOL}^i - t_p)$, is given by the distribution of w_p^i. Figure 4 shows the flow diagram of the prediction process.

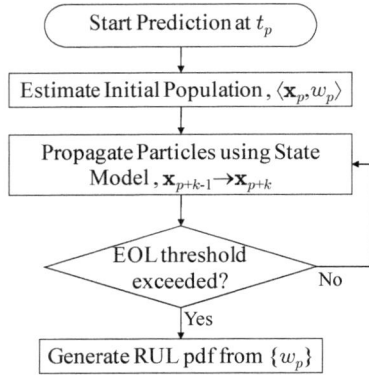

Figure 4. Prediction flowchart.

4. MODEL ADAPTATION

Now that the PF prognostic framework has been set up, let us investigate how we can take advantage of it to perform model adaptation online. For most engineered systems models for nominal operation are available, but true prognostic models like Arrhenius model or Paris' law are comparatively rare. As mentioned before, developing these models require extensive destructive testing which may not be possible in many cases. In some cases, testing may be done on subscale systems, but there may be difficulty in generalizing the models learned. Additionally, the parameter values of these models are often system specific, and thus need to be re-learned for every new application. The PF framework described above can help in these cases by adapting the prognostic/aging model in an online fashion.

For the purposes of this paper we shall assume that the system health state is 1-dimensional, given by x_k, and the state evolution model $\mathbf{f}$ and the measurement model $\mathbf{h}$ are stationary in nature with known noise distributions ω and v respectively. Additionally, we also assume that the parameter values of $\mathbf{h}$ are known. This assumption can be relaxed in a more generic approach. Indeed, considering a non-stationary measurement model can be used to account for progressive degradation in sensors caused by corrosion, fatigue, wear, etc. The parameters of $\mathbf{f}$, denoted by $\alpha_k = \{\alpha_{j,k}; \ j = 1,...,n_f\}$, $n_f \in \mathbb{N}$, are combined with x_k to give the state vector $\mathbf{x}_k = [x_k \ \alpha_k]^T$, where T represents the transpose of a vector or matrix. Equations (11) and (12) can then be rewritten as:

$$x_k = \mathbf{f}(x_{k-1}, \alpha_{k-1}) + \omega_{k-1} \qquad (18)$$

$$z_k = \mathbf{h}(x_k) + v_k. \qquad (19)$$

The issue now is to formulate the state equations for α_k. One easy solution is to pick a *Gaussian random walk* such that:

$$\alpha_{j,k} = \alpha_{j,k-1} + \omega_{j,k-1} \qquad (20)$$

where $\omega_{j,k-1}$ is drawn from a normal distribution, $\mathcal{N}(0, \sigma_j^2)$, with zero mean and variance σ_j^2. Given a suitable starting point $\alpha_{j,0}$, and variance σ_j^2, the PF estimate will converge to the actual parameter value $\bar{\alpha}_j$, according to the *law of large numbers*. In this way, we appear to have introduced model adaptation into the PF framework, adding n_f extra dimensions, yet achieving convergence without incurring the curse of dimensionality.

The notion of a good proposal density, though, comes into play in the choice of the values of $\alpha_{j,0}$ and σ_j^2. If the initial estimate $\alpha_{j,0}$ is far from the actual value and the variance σ_j^2 is small, then the filter may take a large number of steps to converge, if at all. The variance value may be chosen to be higher in order to cover more state-space, but that can also delay convergence. One way to counter this is to make the noise variance itself a state variable that increases if the associated weight is lower than a preset threshold, i.e., the estimated parameter value is far from the true value, and vice-versa. Equation (20) then may be rewritten as:

$$\alpha_{j,k} = \alpha_{j,k-1} + \omega_{j,k-1}; \quad \omega_{j,k-1} \sim \mathcal{N}(0, \sigma_{j,k-1}^2), \qquad (21)$$

$$\sigma_{j,k} = c_{j,k} \cdot \sigma_{j,k-1}; \begin{cases} c_{j,k} < 1, \text{if } w_{k-1} > w_{\text{th}}, \\ c_{j,k} = 1, \text{if } w_{k-1} = w_{\text{th}}, \\ c_{j,k} > 1, \text{if } w_{k-1} < w_{\text{th}}. \end{cases} \qquad (22)$$

The multiplier $c_{j,k}$, is a positive valued real number, while the threshold w_{th} is some value in the interval (0, 1). The intent is to increase the search space when the error is high and tightening the search when we are close to the target. Note that although this produces a better proposal density, it introduces a further n_f dimensions to the state vector.

5. SENSITIVITY ANALYSIS

It is quickly evident that it is not feasible to take this approach for all the parameters of a sufficiently high-order model. This motivates the use of sensitivity analysis techniques (SA) to determine the more sensitive parameters that need to be estimated online.

SA is essentially a methodology for systematically changing parameters in a model to determine the effects on the model output. There are several methods to perform SA like local derivatives (Cacuci, 2003), sampling (Helton *et al.*, 2006), Monte Carlo sampling (Saltelli et al., 2004), etc. Depending upon the form of the system model any of these methods may be used assess which parameters to target.

In this paper, we assume that the model function $\mathbf{f}$ in Eq. (18) is differentiable, i.e., we can compute $\partial \mathbf{f}/\partial \alpha_j$, time index k dropped for the sake of generality, at any point in the state space defined by $\mathbf{x}_k = [x_k \ \alpha_k]^T$. If the partial derivative is positive, then the value of the function increases with an increase in the parameter value and vice-versa. The magnitude of the derivative indicates the degree to which the parameter affects the output of $\mathbf{f}$. This allows us to choose the parameters to estimate online. For example consider the function:

$$\mathbf{f}(x) = \alpha_1 . \exp(\alpha_2 x) \qquad (23)$$

where α_1 and α_2 are the function parameters. Then the partial derivatives are given by:

$$\frac{\partial \mathbf{f}}{\partial \alpha_1} = \exp(\alpha_2 x), \qquad (24)$$

$$\frac{\partial \mathbf{f}}{\partial \alpha_2} = \alpha_1 x . \exp(\alpha_2 x). \qquad (25)$$

Figure 5 shows the sensitivity analysis of $\mathbf{f}(x)$ due to 10% variation in parameters α_1 and α_2 around the value 10, with $x = 1$.

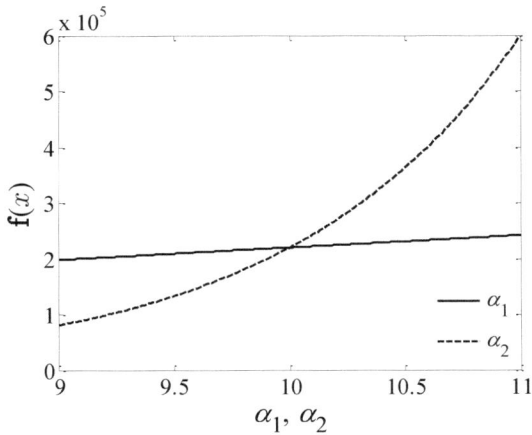

Figure 5. Effect on $\mathbf{f}(x)$ due to 10% variation in parameters α_1 and α_2.

As expected in this simple example, the output of the function is more sensitive to similar variations in the exponential coefficient α_2 than the multiplier α_1, almost by an order of magnitude. Depending on the desired estimation accuracy, α_2 makes a better candidate for online identification than α_1.

Another possibility to note is to replace the random walk model for parameter identification by one that takes into account how a change in the parameter value affects the model output. A similar concept has been applied by Orchard et al., (2009), where they incorporate information from the short term prediction error back into the estimation routine to improve PF performance for both state estimation and prediction. In the case of our example we can construct a similar framework by considering the posterior state error:

$$e_k^i = x_k^i - \sum_{i=1}^{N} w_k^i x_k^i . \qquad (26)$$

If e_k^i is positive then the parameters that have a positive local partial derivative need to be reduced and those with a negative one need to be increased. The opposite holds true if e_k^i is negative. The amount by which the parameters need to be reduced or increased also depends on the magnitude of the local partial derivative. The higher the magnitude, the smaller steps we take in order to prevent instability while approaching the true value. We can formalize this notion in the following way (the particle index i has been dropped for the sake of generality):

$$\alpha_{j,k} = \alpha_{j,k-1} + C_{j,k} + \omega_{j,k-1}; \quad \omega_{j,k-1} \sim \mathcal{N}\left(0, \sigma_j^2\right). \qquad (27)$$

$$C_{j,k} \propto -e_k ,$$

$$\propto \left. \frac{\partial \mathbf{f}}{\partial \alpha_{j,k}} \right|_{\mathbf{x}_k} ,$$

$$= -K . \frac{e_k}{\partial \mathbf{f}/\partial \alpha_{j,k} \big|_{\mathbf{x}_k}} . \qquad (28)$$

Note that in this model adaptation scenario we are not adding the noise variance parameter to the state vector since the search process is directed and not random as discussed in the precious section.

6. PREDICTING BATTERY DISCHARGE

The application example chosen to investigate the notions described above is the discharge of Lithium-ion rechargeable batteries. The electro-chemistry behind the process as well as the model derivation has been discussed in detail in (Saha & Goebel, 2009). Some information is repeated here to maintain readability. For the empirical charge depletion model considered here, we express the output voltage $E(t_k)$ of the cell in terms of the effects of the changes in the internal parameters, as shown below:

$$E(t_k) = E^\circ - \Delta E_{sd}(t_k) - \Delta E_{rd}(t_k) - \Delta E_{mt}(t_k) \quad (29)$$

where E° is the Gibb's free energy of the cell, ΔE_{sd} is the drop due to self-discharge, ΔE_{rd} is the drop due to cell reactant depletion and ΔE_{mt} denotes the voltage drop due to internal resistance to mass transfer (diffusion of ions). These individual effects are modeled as:

$$\Delta E_{sd}(t_k) = \alpha_{1,k} \cdot \exp(-\alpha_{2,k}/t_k), \quad (30)$$

$$\Delta E_{rd}(t_k) = \alpha_{3,k} \cdot \exp(\alpha_{4,k}t_k), \quad (31)$$

$$\Delta E_{mt}(t_k) = \Delta E_{init} - \alpha_{5,k}t_k. \quad (32)$$

where ΔE_{init} is the initial voltage drop when current flows through the internal resistance of the cell at the start of the discharge cycle, and $\alpha_k = \{\alpha_{j,k}; j = 1,...,5\}$ represents the set of model parameters to be estimated. Figure 6 shows how the different voltage drop components defined in Eqns. (30)–(32) combine to give the typical constant current Li-ion discharge profile.

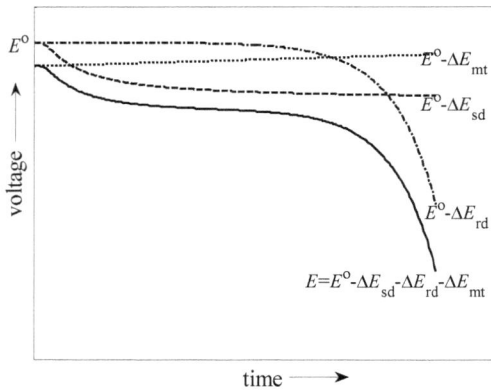

Figure 6. Decomposition of the Li-ion discharge profile in to different components (Saha & Goebel, 2009).

The problem is to predict the end-of-discharge (EOD), i.e., the time instant t_{EOD} when the state x denoting the cell voltage E reaches the threshold level of 2.7 V. The PF representation of this problem is given by:

$$x_k = x_{k-1} - \{\alpha_{1,k-1}\alpha_{2,k-1}\exp(-\alpha_{2,k-1}/t_{k-1})t_{k-1}^2 \\ - \alpha_{3,k-1}\alpha_{4,k-1}\exp(\alpha_{4,k-1}t_{k-1}) - \alpha_{5,k-1}\}(t_k - t_{k-1}) \quad (33) \\ + \omega_{k-1}$$

$$z_k = x_k + v_k. \quad (34)$$

This is a 6 dimensional state vector with 1 dimension being the system health indicator (cell voltage) and the other dimensions coming from the model parameters.

This is a sufficiently complex problem to investigate the PF-based model adaptation techniques described in the paper, since the critical health variable, battery voltage, is dependent on multiple simultaneous internal processes that are not independently observable. Additionally, the voltage undergoes a very steep and nonlinear transformation near the EOD threshold, as shown in Figure 6, which is difficult to predict early on. For simple voltage tracking purposes, a *random walk* model over the cell voltage, i.e. $E(t_{k+1}) = E(t_k) + \omega_{k-1}$, is enough, but when the voltage trajectory needs to be predicted on the basis of present estimates, then accurate estimates of the underlying model parameters are indispensible. This point is illustrated in Figure 7, which shows that a 10% error in estimating the model parameters $\{\alpha_j; j = 1,...,5\}$ can lead to a 15 minute error in determining the remaining battery life.

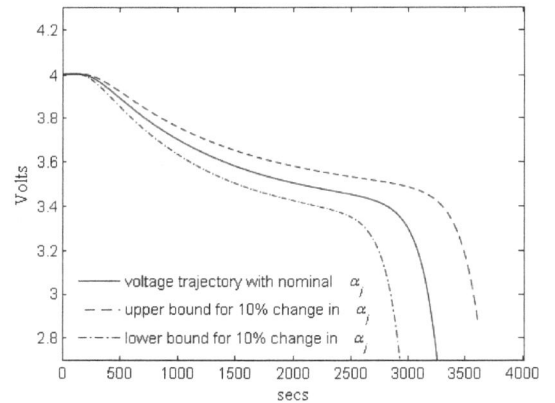

Figure 7. Li-ion discharge trajectories with changes in model parameter estimates.

7. RESULTS

The suitability of using the proposed model adaptation routines, described in Sections 4 and 5, for EOD prediction is measured using the α–λ metric defined in (Saxena et al., 2008). Multiple predictions are made as the battery progressively discharges at a constant current of 2 A. The data have been collected from a custom built battery prognostics testbed at the NASA Ames Prognostics Center of Excellence (PCoE). An example of the PF prediction output based on 50 particles is shown in Figure 8. The prediction points are denoted by stars in blue. The EOD pdfs overlap as shown on the bottom right with the earlier predictions more faded than the newer ones.

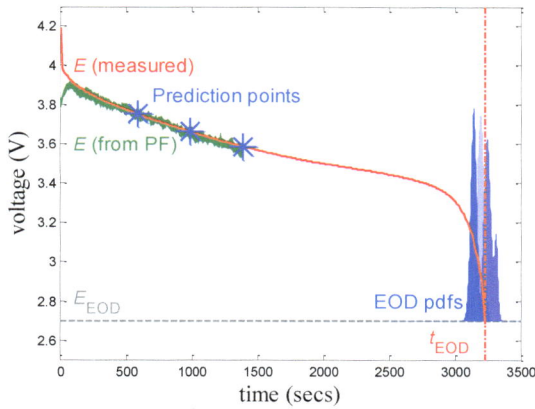

Figure 8. EOD prediction (Saha & Goebel, 2009).

Three different model adaptation routines have been tried:

- *Type A* – the parameters are adapted according to the Gaussian random walk model described in Eq. (20).

- *Type B* – the parameters are adapted based on the noise variance variation strategy described in Eqs. (21) and (22). The threshold w_{th} is chosen to be 0.5.

- *Type C* – the parameters are adapted according to the sensitivity analysis based strategy described in Eqs. (27) and (28). The proportionality factor K is chosen to be 10^5.

For each type of model adaptation 10 EOD prediction runs are conducted each including 13 predictions performed at predetermined time instants. The number of particles is 50 in all cases. The initial population $\langle \mathbf{x}_0^i, w_0^i \rangle$ is also the same for all runs, with $w_0^i = 1/50$. The initial values of the parameters have been learned from discharge runs at 4 A in order to test the model adaptation performance. Figure 9 shows an example of the variation in parameter values at different discharge levels.

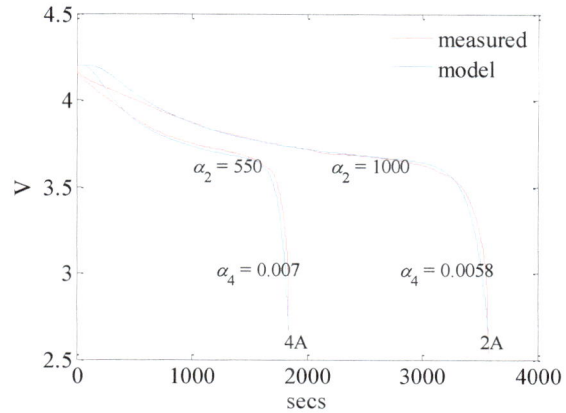

Figure 9. Difference in parameter values for different load currents.

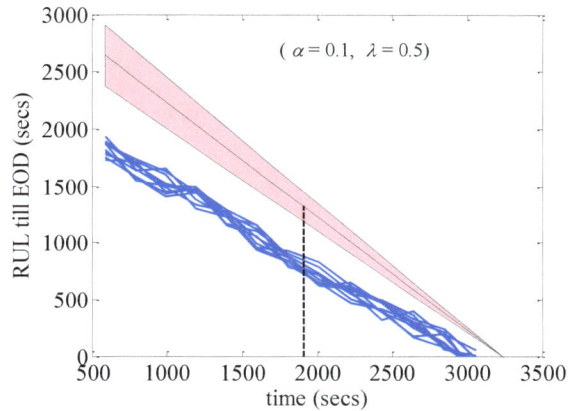

Figure 10. Prognostic performance of model adaptation *type A*.

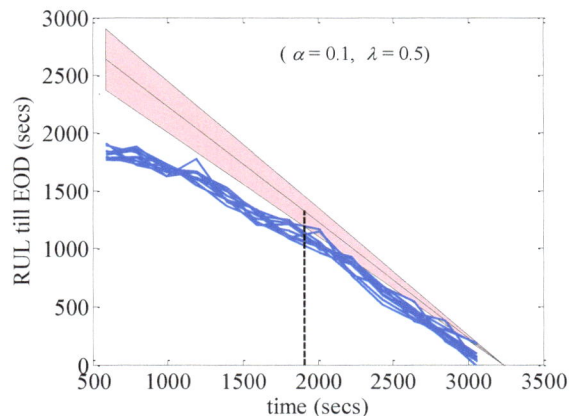

Figure 11. Prognostic performance of model adaptation *type B*.

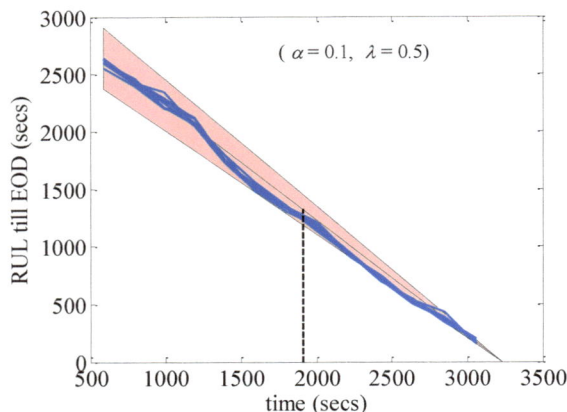

Figure 12. Prognostic performance of model adaptation *type C*.

Figures 10 – 12 show the prognostic performance of the 10 prediction runs of each model adaptation type. As can be seen from Figure 10 the noise variance selected for the model in insufficient to overcome the error between the initial parameter population and the true value.

Figure 11 shows that the noise variance adaptation routine is capable of achieving convergence although it takes up almost half of RUL from the point of prediction to EOD. The SA based adaptation routine performs the best with convergence within 10% (α = 0.1) throughout the prediction horizon as shown in Figure 12, i.e., the model adaptation takes place within the first 500 secs of the discharge. The multiple runs allow us to have some statistical confidence in these results.

Overall, if prognostic performance is evaluated at the 50% mark of the full prediction horizon (λ = 0.5) then only *type C* meets the 10% error performance criterion. In the context of decision making, this prediction can be used to take corrective actions with more than 20 mins remaining. For battery applications, such corrective actions could include altering the load to match the desired battery life.

8. CONCLUSION

In summary, this paper investigates the possibility of performing model adaptation in a PF framework without incurring the curse of dimensionality. It has been shown how various strategies may be used to adapt model parameters online in order to tune the state model for RUL predictions. The feasibility of doing this without incurring the curse of dimensionality has

been demonstrated by the application of sensitivity analysis techniques.

However, the analysis performed in this paper is still preliminary in nature since the effects of the initial populations and the priors chosen for the noise variances have not been investigated. Additionally, theoretical analysis of PF convergence bounds while using model adaptation techniques is necessary for the adoption of these methods into Prognostic Health Management (PHM) practice, and will be tackled in future papers.

ACKNOWLEDGEMENT

The funding for this work was provided by the NASA Integrated Vehicle Health Management (IVHM) project under the Aviation Safety Program of the Aeronautics Research Mission Directorate (ARMD).

REFERENCES

Bellman, R.E. (1957). *Dynamic Programming*, Princeton University Press, Princeton, NJ.

Cacuci, D. G. (2003). *Sensitivity and Uncertainty Analysis: Theory, Volume I*, Chapman & Hall.

Daum, F. E. (2005). Nonlinear Filters: Beyond the Kalman Filter, *IEEE A&E Systems Magazine*, vol. 20, no. 8, pp. 57-69.

Daum, F. E. & Huang, J. (2003). Curse of Dimensionality and Particle Filters, *in Proceedings of IEEE Conference on Aerospace*, Big Sky, MT.

Gordon, N. J., Salmond, D. J. & Smith, A. F. M. (1993). Novel Approach to Nonlinear/Non-Gaussian Bayesian State Estimation, *Radar and Signal Processing, IEE Proceedings F*, vol. 140, no. 2, pp. 107-113.

Helton, J. C., Johnson, J. D., Salaberry, C. J. & Storlie, C. B. (2006). Survey of sampling based methods for uncertainty and sensitivity analysis, *Reliability Engineering and System Safety*, vol. 91, pp. 1175–1209.

Jazwinski, A. H. (1970). *Stochastic Processes and Filtering Theory*, Academic Press, N. Y.

Orchard, M., Tobar, F. & Vachtsevanos, G.. (2009). Outer Feedback Correction Loops in Particle Filtering-based Prognostic Algorithms: Statistical Performance Comparison, *Studies in Informatics and Control*, vol. 18, issue 4, pp. 295-304.

Ristic, B., Arulampalam, S. & Gordon, N. (2004). *Beyond the Kalman Filter*, Artech House.

Saha, B. & Goebel, K. (2009). Modeling Li-ion Battery Capacity Depletion in a Particle Filtering Framework, *in Proceedings of the Annual Conference of the Prognostics and Health Management Society 2009*, San Diego, CA.

Saha, B., Goebel, K., Poll, S. & Christophersen, J. (2009). Prognostics Methods for Battery Health Monitoring Using a Bayesian Framework, *IEEE Transactions on Instrumentation and Measurement,* vol.58, no.2, pp. 291-296.

Saltelli, A., Tarantola, S., Campolongo, F. & Ratto, M. (2004). *Sensitivity Analysis in Practice: A Guide to Assessing Scientific Models*, John Wiley and Sons.

Saxena, A., Celaya, J., Balaban, E., Goebel, K., Saha, B., Saha, S. & Schwabacher, M. (2008). Metrics for Evaluating Performance of Prognostic Techniques, *in Proceedings of Intl. Conf. on Prognostics and Health Management*, Denver, CO.

Bhaskar Saha received his Ph.D. from the School of Electrical and Computer Engineering at Georgia Institute of Technology, Atlanta, GA, USA in 2008. He received his M.S. also from the same school and his B. Tech. (Bachelor of Technology) degree from the Department of Electrical Engineering, Indian Institute of Technology, Kharagpur, India. He is currently a Research Scientist with Mission Critical Technologies at the Prognostics Center of Excellence, NASA Ames Research Center. His research is focused on applying various classification, regression and state estimation techniques for predicting remaining useful life of systems and their components, as well as developing hardware-in-the-loop testbeds and prognostic metrics to evaluate their performance. He has been an IEEE member since 2008 and has published several papers on these topics.

Kai Goebel received the degree of Diplom-Ingenieur from the Technische Universität München, Germany in 1990. He received the M.S. and Ph.D. from the University of California at Berkeley in 1993 and 1996, respectively. Dr. Goebel is a senior scientist at NASA Ames Research Center where he leads the Diagnostics & Prognostics groups in the Intelligent Systems division. In addition, he directs the Prognostics Center of Excellence and he is the Associate Principal Investigator for Prognostics of NASA's Integrated Vehicle Health Management Program. He worked at General Electric's Corporate Research Center in Niskayuna, NY from 1997 to 2006 as a senior research scientist. He has carried out applied research in the areas of artificial intelligence, soft computing, and information fusion. His research interest lies in advancing these techniques for real time monitoring, diagnostics, and prognostics. He holds eleven patents and has published more than 100 papers in the area of systems health management.

INTERNATIONAL JOURNAL OF PROGNOSTICS AND HEALTH MANAGEMENT, VOL.2 (2011)

Applying the General Path Model to Estimation of Remaining Useful Life

Jamie Coble[1] and J. Wesley Hines[2]

[1,2]*The University of Tennessee, Knoxville, TN 37996, USA*

jcoble1@utk.edu
jhines2@utk.edu

ABSTRACT

The ultimate goal of most prognostic systems is accurate prediction of the remaining useful life of individual systems or components based on their use and performance. This class of prognostic algorithms is termed effects-based, or Type III, prognostics. A unit-specific prognostic model, called the General Path Model, involve identifying an appropriate degradation measure to characterize the system's progression to failure. A functional fit of this parameter is then extrapolated to a pre-defined failure threshold to estimate the remaining useful life of the system or component. This paper proposes a specific formulation of the General Path Model with dynamic Bayesian updating as one effects-based prognostic algorithm. The method is illustrated with an application to the prognostics challenge problem posed at PHM '08.

1. INTRODUCTION

Prognostics is a term given to equipment life prediction techniques and may be thought of as the "holy grail" of condition based maintenance. Prognostics can play an important role in increasing safety, reducing downtime, and improving mission readiness and completion. Prognostics is one component in a full health management system (Figure 1). Health monitoring systems commonly employ several modules, including but not limited to: system monitoring, fault detection, fault diagnostics, prognostics, and management (Kothamasu et al., 2006 and Callan et al., 2006). System monitoring and fault detection modules are used to determine if a component or system is operating in a nominal and expected way. If a fault or anomaly is detected by the monitoring system, the diagnostic system determines the type, and in some cases, the severity of the fault. The prognostics module uses this

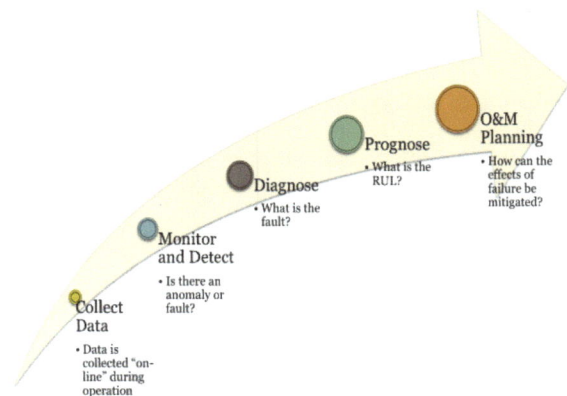

Figure 1: A Full Prognostics and Health Management System

information to estimate the Remaining Useful Life (RUL) of the system or component along with associated confidence bounds. With this information in hand, system operation may be adjusted to mitigate the effects of failure or to slow the progression of failure, thereby extending the RUL to some later point, such as a previously scheduled maintenance activity or the end of the planned mission.

Prognostic system development has been a daunting task for several reasons. One is that mission critical systems are rarely allowed to run to failure once degradation has been detected. This makes the existence of degradation data rare and the development of degradation based models difficult. However, current individual-based, empirical prognostic techniques necessitate the availability of a population of exemplar degradation paths for each fault mode of interest. In some cases, physical models may be developed to generate simulated degradation data or may be used in a model-based prognostics framework to infer RUL (Pecht and

Dasgupta, 1995, Valentin et al., 2003, and Oja et al. 2007). Second, if the components are subject to common fault modes which lead to failure, these fault modes are often designed out of the system through a proactive continuous improvement process. Third, very few legacy systems have the instrumentation required for accurate prognostics. In the absence of such instrumentation, accurate physics of failure models may be used to identify key measurements and systems may be re-instrumented.

This research focuses on RUL estimation for soft failures. These failures are considered to occur when the degradation level of a system reaches some predefined critical failure threshold, e.g. light output from fluorescent light bulbs decreases below a minimum acceptable level or car tire tread is thinner than some pre-specified depth. These failures generally do not concur with complete loss of functionality; instead, they correspond with the time when an operator is no longer confident that equipment will continue to work to its specifications.

Traditional reliability analysis, termed Type I prognostics, uses only failure time data to estimate a time to failure distribution (Hines et al., 2007). This class of algorithms characterizes the average lifetime of an average component operating in historically average conditions; it does not consider any unit-specific information beyond the current run time. As components become more reliable, few failure times may be available, even with accelerated life testing. Although failure time data become more sporadic as equipment reliability rises, often other measures are available which may contain some information about equipment degradation, such as crack length, tire pressure, or pipe wall thickness. Lu and Meeker (1993) developed the General Path Model (GPM) to model equipment reliability using these degradation measures, or appropriate functions thereof, moving reliability analysis from failure-time analysis to failure-process analysis. The GPM assumes that there is some underlying parametric model to describe component degradation. The model may be derived from physical models or from available historical degradation data. Typically, this model accounts for both population (fixed) effects and individual (random) effects.

Although GPM was originally conceived as a method for estimating population reliability characteristics, such as the failure time distribution, it has since been extended to individual prognostic applications (Upadhyaya et al., 1994). Most commonly, the fitted model is extrapolated to some known failure threshold to estimate the RUL of a particular component. This is an example of an Effects-based, or Type III, prognostic algorithm (Hines et al., 2007). This class of algorithms estimates the RUL of a specific component or system operating in its specific environment; it is the ultimate goal of prognostics for most mission critical components.

The following sections will present GPM theory including the original methodology for reliability applications and the extension to prognostics. In addition, a short discussion of dynamic Bayesian updating methods to incorporate prior information is given. Finally, an application of the proposed GPM methodology to the 2008 PHM Challenge problem is presented.

2. METHODOLOGY

As suggested by the "No Free Lunch" Theorem, no one prognostic algorithm is ideal for every situation (Koppen, 2004). A variety of models have been developed for application to specific situations or specific classes of systems. The efficacy of these algorithms for a new process depends on the type and quality of data available, the assumptions inherent in the algorithm, and the assumptions which can validly be made about the system. This research focuses on the general path model, an algorithm which attempts to characterize the lifetime of a specific component based on measures of degradation collected or inferred from the system.

2.1. The General Path Model

Lu and Meeker (1993) first proposed the General Path Model (GPM), an example of degradation modeling, to move reliability analysis methods from time-of-failure analysis to process-of-failure analysis. Traditional methods of reliability estimation use failure times recorded during normal use or accelerated testing to estimate a time of failure (TOF) distribution for a population of identical components. In contrast, GPM uses degradation measures to estimate the TOF distribution. The use of historical degradation measures allows for the direct inclusion of censored data, which gives additional information on unit-wise variations in a population.

GPM analysis begins with some assumption of an underlying functional form of the degradation path for a specific fault mode. The degradation of the ith unit at time t_j is given by:

$$y_{ij} = \eta\left(t_j, \varphi, \theta_i\right) + \varepsilon_{ij} \qquad (1)$$

where φ is a vector of fixed (population) effects, θ_i is a vector of random (individual) effects for the ith component, and $\varepsilon_{ij} \sim N(0,\sigma^2_\varepsilon)$ is the standard measurement error term. Application of the GPM methodology involves several assumptions. First, the degradation data must be describable by a function, η; this function may be derived from physics-of-failure models or from the degradation data itself. In order to fit this model, the second assumption is that historical degradation data from a population of identical components or systems are available or can be simulated. This data should be collected under similar use (or accelerated test) conditions and should reasonably span the range of individual variations between components. Because GPM uses degradation measures instead of failure times, it is also not necessary that all historical units are run to failure; censored data contain information useful to GPM forecasting. The final assumption of the GPM model is that there exists some defined critical level of degradation, D, which indicates component failure; this is the point beyond which the component will no longer perform its intended function with an acceptable level of reliability. Therefore, some components should be run to failure, or to a state considered failure, in order to quantify this degradation level. Alternatively, engineering judgment may be used if the nature of the degradation parameter is explicitly known.

Several methods are available to estimate the degradation model parameters, φ and θ. In some cases, the population parameters may be known in advance, such as the initial level of degradation. If the population parameters are unknown, estimation of the vector of population characteristics, φ, is trivial; by fitting the model to each exemplar degradation path, the fixed effects parameters can be taken as the mean of the fitted values for each unit. The variance of these estimates should be examined to ensure that the parameters can be considered to be fixed. If significant variability is present, the parameters should be considered random and moved to the θ vector. A two-stage method of parameter estimation was proposed by Lu and Meeker (1993) to estimate distribution parameters for the random effects.

In the first stage, the degradation model is fit to each degradation path to obtain an estimate of θ for that unit; these θ's are referred to as stage-1 estimates. It is convenient to assume that the stage-1 estimates, or an appropriate transformation, $\Theta=H(\theta)$, is normally (or multivariate normally) distributed so that the random effects can be fully described using only a mean vector and variance-covariance matrix without significant loss of information. This assumption usually holds for large populations as a result of the central limit theorem; however, if it is not justifiable, the GPM methodology can be extended in a natural way to allow for other random effects distributions.

In the second stage, the stage-1 estimates (or an appropriate transformation thereof) are combined to estimate φ, μ_θ, and Σ_θ. At this stage, if for any random parameter, m, the variance σ^2_m is effectively zero, this parameter should be considered a fixed effects parameter and should be removed from the random parameter distribution.

In their seminal paper, Lu and Meeker (1993) describe Monte Carlo methods for using the GPM parameter estimates to estimate a time to failure distribution and corresponding confidence intervals. Because the focus of this paper is estimating time to failure of an individual component and not the failure time distribution of the population of components, these methods will not be described here.

Several limitations and areas of future work of the GPM are identified by Meeker et al. (1998). Some of these areas have been addressed in work by other authors. First, the authors cite the need for more accurate physics of failure models. While such models are helpful for understanding degradation mechanisms, they may not be strictly necessary for RUL estimation. In fact, if exemplar data sets cover the range of likely degradation paths, it may be adequate to fit a function which does not explain failure modes but accurately models the underlying relationships. With this idea, neural networks have been applied to GPM reliability analysis (Chinnam, 1999 and Girish et al., 2003).

In addition, the GPM was originally developed for reliability analysis of only one fault mode. In practical applications, the system of interest may consist of several components each with different fault modes, or of one component with several possible, even simultaneous fault modes. These multiple degradation paths may be uncorrelated, in which case extension of the GPM is trivial: reliability of a component for all degradation modes is simply the product of the individual reliabilities, and RUL can be considered some function of the RULs for each fault mode, such as the minimum. If, however, the degradation measures are correlated, extension of the GPM is more complicated. For example, in the case of tire monitoring, several degradation measures may contain information about tire reliability, including tread thickness, tire pressure, tire temperature and wall material characteristics. However, it is easy to see that these measures may be correlated; a higher temperature would cause a higher pressure etc. The case of multiple, competing degradation modes is beyond the scope of the current work. A discussion of the problem can be found in Wang and Coit (2004).

2.2. GPM for Prognostics

The GPM reliability methodology has a natural extension to estimation of remaining useful life of an individual component or system; the degradation path model, y_i, can be extrapolated to the failure threshold, D, to estimate the component's time of failure. This type of degradation extrapolation was proposed early on by Upadhyaya et al. (1994). In that work, the authors used both neural networks and nonlinear regression models to predict the RUL of an induction motor. The prognostic methodology used for the current research is described below.

First, exemplar degradation paths are used to fit the assumed model. The stage-1 parameter estimates are used to evaluate the random-effects distributions, to determine the mean population random effects, the mean time to failure (MTTF) and their associated standard deviations, and to estimate the noise variance in the degradation paths. The MTTF distribution can be used to estimate the time of failure for any component which has not yet been degraded.

As data are collected during use, the degradation model can be fit for the individual component. This specific model can be used to project a time of failure for the component. Because of noise in the degradation signal, the projected time of failure is not perfect. A prediction interval (PI) about the estimated parameters can be evaluated as:

$$\theta \in \left[\hat{\theta} - t_{n-1,\frac{\alpha}{2}} s \sqrt{1 + \frac{1}{n}}, \ \hat{\theta} + t_{n-1,\frac{\alpha}{2}} s \sqrt{1 + \frac{1}{n}} \right] \quad (2)$$

where $t_{n-1,\alpha/2}$ is the Student's t-distribution, n is the number of observations used to fit the model, and s is the standard deviation of the degradation model parameters for normally distributed, uncorrelated parameters; if this assumption is not met, the method can be extended to estimate PIs for other distributions. The standard deviation of the parameters can be estimated through traditional linear regression techniques. The range of model parameters can be used to project an PI about the estimated time of failure.

The methodology described considers only the data collected on the current unit to fit the degradation model. However, prior information available from historic degradation paths can be used for initial model fitting, including the mean degradation path and associated distributions. This data can provide valuable knowledge for fitting the degradation model of an individual component, particularly when only a few data points have been collected or the collected data suffer from excessive noise. The following section outlines a dynamic Bayesian updating method for including prior information in degradation model fitting.

2.3. Incorporating Prior Information

The current research investigates using Bayesian methods to include prior information for linear regression problems. However, as discussed above, the GPM methodology can be applied to nonlinear regression problems as well as

other parametric modeling techniques such as neural networks. Other Bayesian methods must be applied to these types of models, but such application is beyond the scope of the current research. For a complete discussion of Bayesian statistics including other Bayesian update methods, the interested reader is referred to Carlin and Louis (2000) and Gelman et al. (2004). In addition, work by Robinson and Crowder (2000) focuses on Bayesian methods for nonlinear regression reliability models.

A brief review of Bayesian update methods for linear regression is given here; a more complete discussion can be found in Lindely and Smith (1972) as well as the texts cited above. Bayesian updating is a method for combining prior information about the set of model parameters with new data observations to give a posterior distribution of the model parameters (Figure 2). This allows both current observation and past knowledge to be considered in model fitting.

A linear regression model is given by:

$$Y = bX \quad (3)$$

The model parameters are estimated using the pseudo-inverse formula as:

$$b = \left(X^T \Sigma_y^{-1} X \right)^{-1} X^T \Sigma_y^{-1} Y$$

$$Y = \begin{bmatrix} y_1 \\ y_2 \\ \vdots \\ y_m \end{bmatrix} \quad X = \begin{bmatrix} x_{11} & x_{12} & \cdots & x_{1n} \\ x_{21} & x_{22} & \cdots & x_{2n} \\ \vdots & \vdots & \ddots & \vdots \\ x_{m1} & x_{m2} & \cdots & x_{mn} \end{bmatrix} \quad (4)$$

$$\Sigma_y = \begin{bmatrix} \sigma_{y_1}^2 & 0 & \cdots & 0 \\ 0 & \sigma_{y_2}^2 & \ddots & \vdots \\ \vdots & \ddots & \ddots & 0 \\ 0 & \cdots & 0 & \sigma_{y_m}^2 \end{bmatrix}$$

where Σ_y is the variance-covariance noise matrix, which gives an indication of the accuracy of each entry in the **Y**-vector. It is important to note that the linear regression model is not necessarily a linear model, but is linear-in-parameters. The data matrix

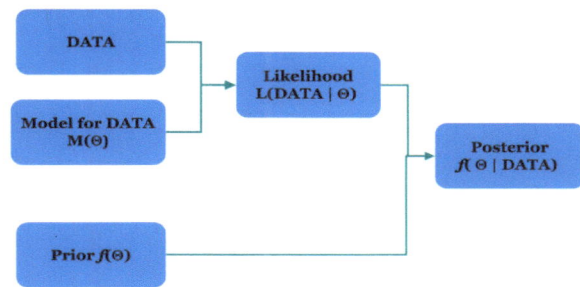

Figure 2: Bayesian Updating Methodology

X can be populated with any function of degradation measures, including higher order terms, interaction terms, and functions such as $sin(x)$ or e^x. If prior information is available for a specific model parameter, i.e. $\beta_j \sim N(\beta_{jo}, \sigma^2_\beta)$, then the matrix **X** should be appended with an additional row with value one at the jth position and zero elsewhere, and the **Y** matrix should be appended with the *a priori* value of the jth parameter.

$$Y^* = \begin{bmatrix} y_1 \\ y_2 \\ \vdots \\ y_m \\ \beta_{j0} \end{bmatrix} \quad X^* = \begin{bmatrix} x_{11} & x_{12} & \cdots & x_{1n} \\ x_{21} & x_{22} & \cdots & x_{2n} \\ \vdots & \vdots & \ddots & \vdots \\ x_{m1} & x_{m2} & \cdots & x_{mn} \\ 0 & \cdots & 1 & 0 \end{bmatrix} \quad (5)$$

Finally, the variance-covariance matrix is augmented with a final row and column of zeros, with the variance of the *a priori* information in the diagonal element.

$$\Sigma^*_y = \begin{bmatrix} \sigma^2_{y_1} & 0 & \cdots & 0 & 0 \\ 0 & \sigma^2_{y_2} & \ddots & \vdots & 0 \\ \vdots & \ddots & \ddots & 0 & \vdots \\ 0 & \cdots & 0 & \sigma^2_{y_m} & 0 \\ 0 & 0 & \cdots & 0 & \sigma^2_{\beta_{j0}} \end{bmatrix} \quad (6)$$

If knowledge is available about multiple regression parameters, the matrices should be appended multiple times with one row for each parameter.

It is convenient to assume that the noise in the degradation measurements is constant and uncorrelated. Some *a priori* knowledge of the noise variance is available from the exemplar degradation paths. If this assumption is not valid for a particular system, then other methods of estimating the noise variance may be used; however, it has been seen anecdotally that violating this assumption does not have a significant impact on RUL estimation. In addition, it is also convenient to assume that the noise measurements are uncorrelated across observations of y; this allows the variance-covariance matrix to be a diagonal matrix consisting of noise variance estimates and *a priori* knowledge variance estimates. If this assumption is not valid, including covariance terms is trivial; again, these terms can be estimated from historical degradation paths.

After *a priori* knowledge is used in conjunction with n current data observations to obtain a posterior estimate of degradation parameters, this estimate becomes the new prior distribution for the next estimation of regression parameters. The variance of this new knowledge is estimated as:

$$\sigma^2_{post} = \left[\frac{1}{\sigma^2_{prior}} + \frac{n}{\sigma^2_y} \right]^{-1} \quad (7)$$

The Bayesian information may be used to dynamically update the model fit as new data become available for each desired RUL estimate.

2.4. Combined Monitoring and Prognostic Systems

Figure 3 shows a combined monitoring, fault detection, and prognostics system similar to the one used in this research. The monitoring system employs an Auto-Associative Kernel Regression (AAKR) model for monitoring and the Sequential Probability Ratio Test (SPRT) for fault detection. Both of these methods are described in broad detail below. The interested reader is referred to (Hines et al., 2008 and Garvey et al., 2007) for a more complete discussion of AAKR and (Wald, 1943) for SPRT.

Auto-Associative models can generally be considered an error correction technique. These models compare a new observation to those seen in the past to estimate how the system "should" be running. These corrected predictions can be compared to the measured data to identify faulted operation. Several auto-associative architectures are available, including auto-associative neural networks, auto-associative kernel regression, and multivariate state estimation technique (Hines et al., 2008). This research employs the AAKR algorithm for system monitoring.

AAKR is a non-parametric, empirical technique. Exemplar historical observations of system operation are stored in a data matrix. As a new observation is collected, it is compared to each of

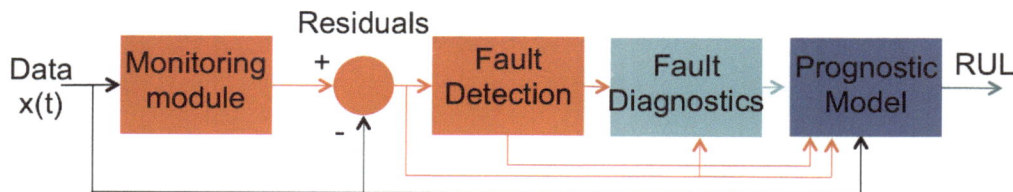

Figure 3: Combined Monitoring and Prognostic System

the exemplar observations to determine how similar the new observation is to each of the exemplars. This similarity is quantified by evaluating the distance between the new observation and the exemplar. Most commonly, the Euclidean distance is used:

$$d_i = \sqrt{\sum_{j=1}^{m} \left(X_j - x_{i,j} \right)^2} \tag{8}$$

where d_i is the distance between the new observation, X, and the ith exemplar, x_i. The distance is converted to a similarity measure through the use of a kernel. Many kernels are available; this research employs the Gaussian kernel:

$$s_i = \exp\left(-\frac{d_i^2}{h^2} \right) \tag{9}$$

where s_i is the similarity of the new observation to the ith exemplar and h is the kernel bandwidth, which controls how close vectors must be to be considered similar. Finally, the "corrected" observation value is calculated as a weighted average of the exemplar observations:

$$\hat{X} = \frac{\sum s_i x_i}{\sum s_i} \tag{10}$$

Monitoring system residuals are then generated as the difference between the actual observation and the error-corrected prediction. These residuals are used with a SPRT to determine if the system is operating in a faulted or nominal condition. As the name suggests, the SPRT looks at a sequence of residuals to determine if the time series of data is more likely from a nominal distribution or a pre-specified faulted distribution. As new observations are made, the SPRT compares the cumulative sum of the log-likelihood ratio:

$$s_i = s_{i-1} + \log \Lambda_i \tag{11}$$

to two thresholds, which depend on the acceptable false positive and false negative fault rates:

$$a \leq \log\left(\frac{\beta}{1-\alpha} \right)$$
$$b \geq \log\left(\frac{1-\beta}{\alpha} \right) \tag{12}$$

where α is the acceptable false alarm (false positive) rate and β is the acceptable missed alarm (false negative) rate. For this research, false alarm and missed alarm rates of 1% and 10% respectively are used. If $s_i < a$, then the null hypothesis cannot be rejected; that is, the system is assumed to be operating in a nominal condition. If $s_i > b$,

then the null hypothesis is rejected; that is, the system is assumed to be operating in a faulted condition. When a determination is made, the sum, s_i, is reset to zero and the test is restarted.

After a fault is detected in the system, the prognostic system can be engaged to determine the RUL for the system. As discussed above, the GPM methodology uses a measure of system degradation, called a prognostic parameter, to make prognostic estimates. An ideal prognostic parameter has three key qualities: monotonicity, prognosability, and trendability.

Monotonicity characterizes the underlying positive or negative trend of the parameter. This is an important feature of a prognostic parameter because it is generally assumed that systems do not undergo self-healing, which would be indicated by a non-monotonic parameter. This assumption is not valid for some components such as batteries, which may experience some degree of self repair during short periods of nonuse, but it tends to hold for mechanical systems or for complex systems as a whole.

Prognosability gives a measure of the variance in the critical failure value of a population of systems. A wide spread in critical failure values can make it difficult to accurately define a critical failure threshold and to extrapolate a prognostic parameter to failure. Prognosability may be very susceptible to noise in the prognostic parameter, but this effect may be reduced by traditional variance reduction methods such as parameter bagging and data denoising.

Finally, trendability indicates the degree to which the parameters of a population of systems have the same underlying shape and can be described by the same functional form.

The population of noise-free prognostic parameters shown in Figure 4 exhibits the three desired features. The parameters are monotonic: they all generally trend upward through time. They are prognosable: the parameter value at failure for each unit is at approximately the same value, as indicated by the red markers. Finally, they are trendable: each parameter appears to follow the same upward exponential or quadratic trend.

Monitoring system residuals, or combinations of residuals, are natural candidates for prognostic parameters because they inherently measure the deviation of a system from normal operation. The following section investigates the application of this monitoring/prognostic method to the 2008 PHM challenge problem.

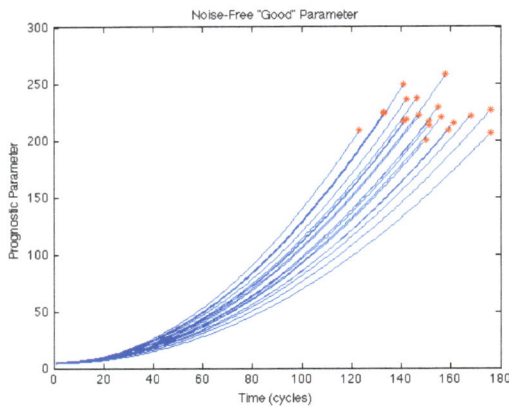

Figure 4: Population of "good" prognostic parameters

3. PHM '08 CHALLENGE APPLICATION

This section presents an application of the proposed GPM prognostic method to the PHM Challenge data set. The efficacy of the method is analyzed based on the given cost function for the 218 test cases. RUL estimates far from the actual value are penalized exponentially. The cost function is asymmetric; RUL predictions greater than the actual value are penalized more heavily than those which predict failure before it happens. The cost for each case is given by the following formula:

$$d = RUL_{estimated} - RUL_{actual}$$
$$score(d < 0) = \exp(-d/13) - 1 \qquad (13)$$
$$score(d > 0) = \exp(d/10) - 1$$

where d is the difference between the estimated and the actual RUL. If d is negative, then the algorithm underestimates the RUL leading one to end operation before failure occurs; if d is positive, then the algorithm overestimates the RUL and results in a greater penalty because one may attempt to operate the component longer than possible and thereby experience a failure. The following sections give a brief description of the simulated data set used for the challenge problem, then outline the data analysis and identification of an appropriate prognostic parameter for GPM trending. Finally, the application of the GPM method and Bayesian updating are presented with final results given for the described method. The performance of the GPM algorithm with and without Bayesian updating is compared.

3.1. PHM Challenge Data Set Description

The PHM Challenge data set consists of 218 cases of multivariate data that track from nominal operation through fault onset to system failure. Data were provided which modeled the damage propagation of aircraft gas turbine engines using the Commercial Modular Aerop-Propulsion System Simulation (C-MAPSS). This engine

simulator allows faults to be injected in any of the five rotating components and gives output responses for 58 sensed engine variables. The PHM Challenge data set included 21 of these 58 output variables as well as three operating condition indicators. Each simulated engine was given some initial level of wear which would be considered within normal limits, and faults were initiated at some random time during the simulation. Fault propagation was assumed to evolve in an exponential way based on common fault propagation models and the results seen in practice. Engine health was determined as the minimum health margin of the rotating equipment, where the health margin was a function of efficiency and flow for that particular component; when this health indicator reached zero, the simulated engine was considered failed. The interested reader is referred to Saxena et al. (2008) for a more complete description of the data simulation.

The data have three operational variables – altitude, Mach number, and TRA – and 21 sensor measurements. Initial data analysis resulted in the identification of six distinct operational settings; based on this result, the operating condition indicators were collapsed into one indicator which fully defined the operating condition of the engine for a single observation (flight). In addition, ten sensed variables were identified whose statistical properties changed through time and were well correlated (linear correlation coefficient of at least 0.7, shown in Figure 5) to each other. In this way, the 24 sensor data set was reduced to 11 variables, with original variable numbers: 1 (the operating condition indicator), 5, 6, 7, 12, 14, 17, 18, 20, 23, and 24.

The GPM method uses degradation information, either directly measured or inferred, to estimate the system RUL. Initial analysis of the raw data does

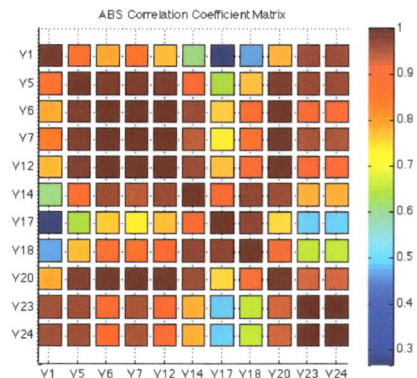

Figure 5: Correlation Coefficient Matrix for Eleven Monitored Variables

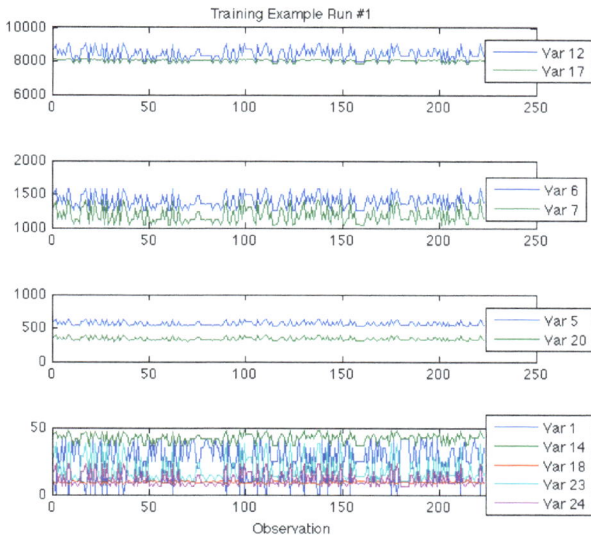

Figure 6: Eleven PHM Data Set Variables

not reveal any trendable degradation parameter. That is, no sensed measurement has an identifiable trend toward failure. Figure 6 is a plot of the eleven variables that were determined to statistically change with time. These variables were used to develop a monitoring and prognostics system. Visual inspection of the data does not indicate any obvious trends toward failure. The monitoring system provides much greater sensitivity to subtle changes that may be indicative of failure.

3.2. Monitoring and Prognostics Results

An AAKR model is used to determine the expected values of the eleven variables of interest. The baseline model is developed using the first 15% of each run as training data; this assumes that faults occur after at least 15% of each run is completed. This assumption is not universally valid, but seems to be reasonable for this data set. Based on the AAKR predictions, a residual is calculated between the nominal prediction and the actual value. These residuals are potential candidates for inclusion into the degradation parameter.

Figure 7 is a plot of sensed variable 17 and the corresponding residual for five of the training cases. The final value for each of the five cases is indicated in the lower plot by red asterisks. In this case, the residual does not provide a useful prognostic parameter. The residual is not trendable; that is, the five cases show several distinct residual shapes. In addition, the residual's prognosability is not high. The residuals end at very different values for each case. This could be indicative of different failure modes, but is not directly useful as a degradation

parameter. For the purpose of this analysis, it is assumed that all fault modes may be lumped together into one prognostic model. Therefore, a single degradation parameter is desired to prognose all systems.

Several of the residuals grow in a similar manner with time for all the units and have failure values without much variation. These residuals can be used as a degradation parameter by trending them through regression and extrapolating the functional fit to some degradation threshold to give an estimate of RUL. The top plot in Figure 8 shows one such sensed variable while the lower plot is the residual between the predicted and measured value. It shows the unfiltered residual for 5 different training cases. This variable is a good prognostic parameter because the corresponding monitoring system residuals of each of the training systems have the same basic shape and failures occur at approximately the same negative value. The task is to model the degradation parameter and predict the failure point when only a subset of the case is given. Five residuals were found to have a similar shape with well clustered values at failure.

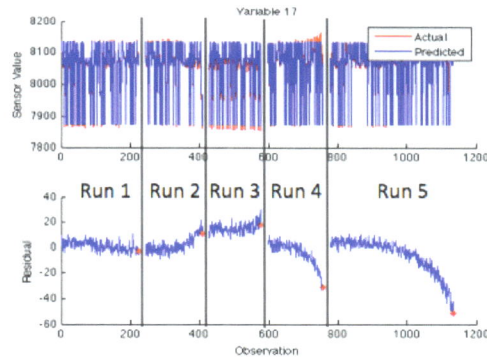

Figure 7: Residual trend indicating possible different failure modes

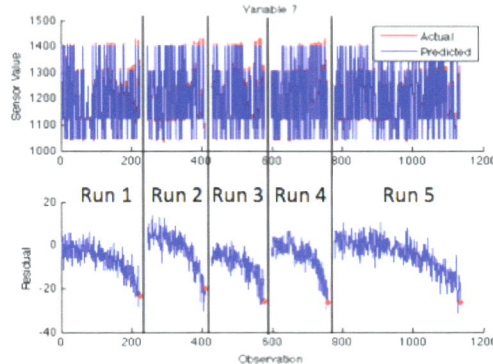

Figure 8: Residual trend candidate for degradation parameter

By combining the five degradation parameters with similar shapes, an average parameter was developed. The five residuals are combined in a weighted average, where each residual weight is inversely proportional to its variance. Figure 9 is a plot of one of the candidate residuals, and Figure 10 is the averaged degradation parameter resulting from a fusion of the five residuals for all 218 cases. As the plots show, the residual parameters have very similar shapes for each training case. However, the single residual is contaminated with greater noise and has a relatively larger spread in the final parameter value. By combining several similar residuals, the spread in the failure value relative to the range of the parameter is significantly reduced, as shown in the second figure. This is sometimes referred to as parameter bagging and is a common variance reduction technique.

A second order polynomial model can been used to model the degradation parameter. While an exponential model may be more physically appropriate, the quadratic model is more robust to noise and better describes the data fit for the chosen prognostic parameter. For the methodology proposed, the model must be linear in parameters; however, simple exponential models, such as $y=exp(ax+b)$ parameterized as $ln(y) = ax +b$, cannot be used with negative y-values, because the natural logarithm of a negative number is undefined in the real number system. This adds unnecessary complexity to the modeling method. Quadratic equations, on the other hand, are naturally linear in parameters and can be used without significant concern for the effects of noise on the model fit. Shifting the prognostic parameter to the positive quadrant by adding 25.0 to every value eliminates the problem of taking the logarithm of negative values; however, the quadratic fit results in a lower fitting error than the exponential fit, with mean squared errors of 1.53 and 2.33 respectively. Because of its robustness to noise and reduced modeling error, the quadratic fit is chosen for this research.

Figure 11 gives an example of a polynomial fit of the prognostic parameter with the time the model crosses the critical failure threshold indicated. The threshold of -13.9 was chosen as the upper 95% level of the distribution of failure values for the known failed cases. This gives an estimated system reliability of 95%, which is a conservative estimate of failure time and reduces the possibility of overestimating RUL leading to in-service failure. The time between the last sample and the estimated time of failure is the estimate of RUL, as indicated by the blue area. For this case, the estimated RUL is exactly correct, with an estimated remaining life of 36 cycles.

The GPM methodology presented works well if many observations are available to fit a model to the degradation parameter as in case 106 shown above. However, when only a few observations have been collected, the model fit is highly susceptible to noise. To counteract this, the Bayesian updating

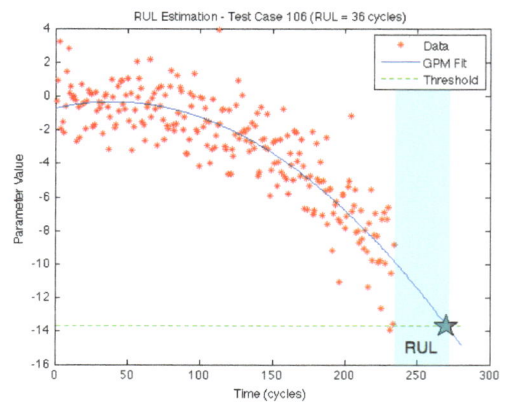

Figure 9: Single residual as a prognostic parameter

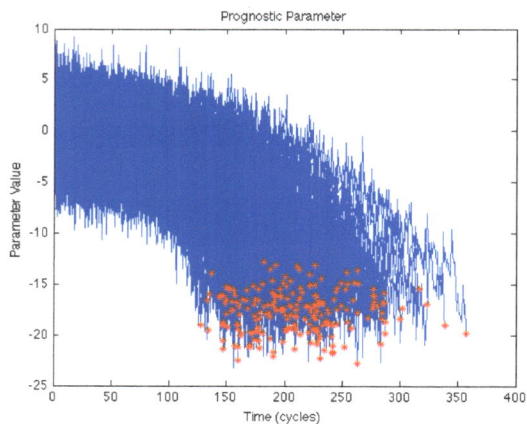

Figure 10: Prognostic parameter for all 218 training cases

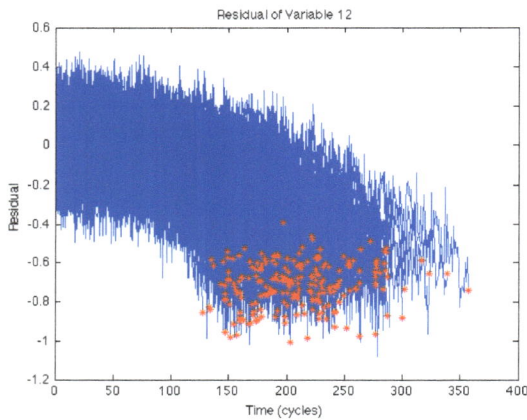

Figure 11: Prognostic parameter trending and RUL estimation

method described previously is used to include prior information about the degradation parameter fit.

For the current problem, quadratic models (eqn. 14) were fit to the full degradation parameter for each of the 218 training cases.

$$d = p_1 t^2 + p_2 t + p_3 \qquad (14)$$

The means and standard deviations for the three parameters (p_i) are given in Table 1. The parameters should be considered random effects because their standard deviations represent a significant proportion of the mean parameter value. The large variance seen in p_3 is assumed to correspond to the random level of initial degradation. The variance of the degradation parameter can be estimated from the training examples by smoothing each example path and subtracting the smoothed path from the actual path. This gives an estimate of the noise; the noise variance can be estimated directly as the variance of this data set. For this data, the noise variance in the degradation parameter is estimated to be 0.0588 units.

	Mean	Std Dev
p1	-0.0001	4.30E-05
p2	0.0075	0.0028
p3	-0.2057	0.37

Table 1: Prior Distribution for Quadratic Parameters

Figure 12 gives an example of a degradation case which is not well fit by the non-Bayesian approach. Few observations (~30% of the total lifetime) are available, and those available have noise levels, which preclude appropriate model fitting. The same data set, fit with the Bayesian approach described and the prior distribution estimates given above is shown in Figure 13. As can be seen, the Bayesian fit reflects the shape seen in the historical degradation paths. The RUL estimate obtained with the Bayesian approach is 135 cycles, versus an undeterminable estimate obtained from the non-Bayesian approach. The actual RUL after the first 84 observations is 170 cycles, resulting in an RUL error of approximately 20%. While this error is still high, it is within a reasonable accuracy considering the amount of data available and will improve as more data are collected.

The advantage of including prior information via dynamic Bayesian updating is to improve RUL estimates when very few observations are available, the data are very noisy, or both. A comparison of the performance through time of the GPM algorithm and the GPM with Bayesian updating, hereafter referred to as GPM/Bayes, is given in Figure 14. In this analysis, the two methodologies were applied to each of the training cases using only a fraction of the full lifetime. The models were applied to subsets of each lifetime in 5% increments, i.e. the models were run using

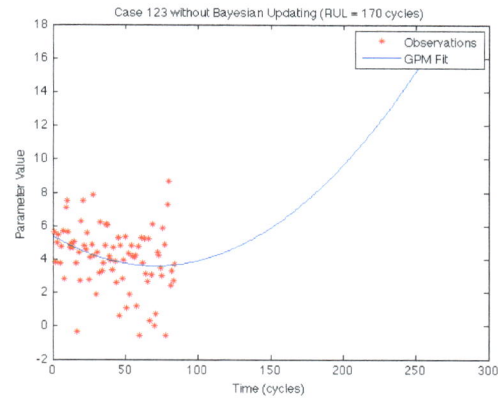

Figure 12: Poor GPM fit

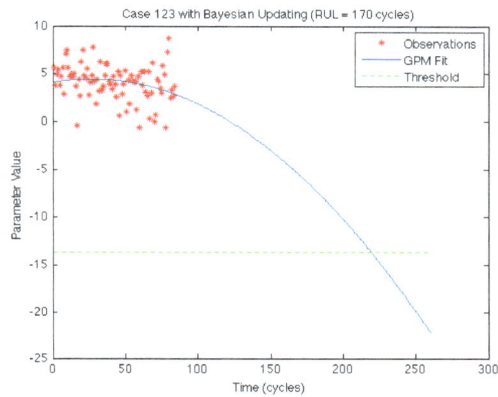

Figure 13: GPM fit with Bayesian update

5% of the full lifetime, 10%, 15%, etc. The RUL error at each percentage was calculated across the 218 full training cases to determine how the error decreases as more data become available. As was seen in the example case above, the non-Bayesian method may result in an undeterminable RUL. In fact, for the data used here, nearly half the runs resulted in an indeterminate RUL estimate using the GPM methodology without Bayesian updating for runs using less than half the total lifetime. For these cases, the RUL is estimated using a Type I, or traditional reliability-based, method in order to give an estimate of RUL prediction error. The mean residual life is found at each time using a Weibull fit of the failure times and the current lifetime (Figure 15). Mean Residual Life (MRL) is found by:

$$MRL(t) = \frac{1}{R(t)} \int_t R(s)\,ds \qquad (15)$$

where $R(t)$ is the reliability function at time t. In practice, the prognostic method would likely fall back to a more rudimentary method such as this if the Type III model did not produce a reasonable answer.

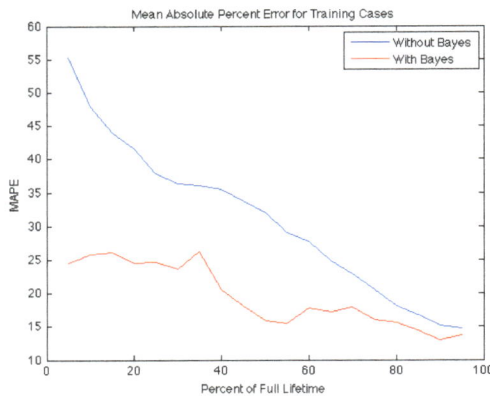

Figure 14: GPM Results With and Without Bayesian Updating

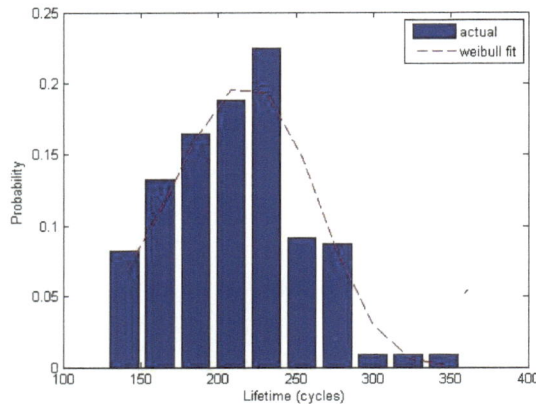

Figure 15: Weibull Probability Fit

The GPM/Type I model which does not include prior information gives an average error of approximately 55% when only 5% of the full lifetime is available and relies on the Type I method for approximately half of the cases. Conversely, the GPM/Bayes method gives approximately 25% error and is able to predict an RUL for every case. As Figure 14 shows, the average error of both methods decreases as more data becomes available and eventually converges to approximately equal error values when the available data overpowers the prior information in the GPM/Bayes model.

4. CONCLUSION

This paper presented a method for performing prognostics on individual components or systems. The General Path Model (GPM) method is used to extrapolate a prognostic parameter curve to a predefined critical failure threshold to obtain an estimate of the Remaining Useful Life (RUL). In cases where only a few data points are available or the data are contaminated by significant noise, a Bayesian method was introduced. The Bayesian method includes prior information about the prognostic parameter

distribution to "force" the functional fit to follow the trend seen in historic systems. The method was applied to the 2008 PHM conference challenge problem to illustrate its efficacy.

The given application utilized the results of a condition monitoring and fault detection system to characterize the degradation in a specific system. A prognostic parameter was generated from a subset of the monitoring system residuals; monitoring system residuals are well-suited components of a prognostic parameter because they naturally characterize the deviation of a system from nominal condition. A parametric, linear-in-parameters regression fit of the time series prognostic parameter was extrapolated to the critical failure threshold to give an estimate of the system RUL. A Bayesian updating method was applied to allow for the inclusions of prior information, which improves model performance particularly when faced with small amounts of data or extremely noisy data. The results show that the GPM/Bayes method greatly improved RUL predictive performance over a conventional regression solution.

The need for a diversity of algorithms suggests that development of a large variety of prognostic methods can only strengthen the field. While the algorithm described here may not be the best performing method for this data set, it has several key advantages that the winning PHM Challenge algorithms lack, which may make it better suited for other applications. The proposed GPM/Bayes algorithm is qualitatively compared to the three best performing algorithms at the 2008 PHM Challenge in the following discussion.

The similarity-based approach described in (Wang et al., 2008) shares several assumptions with the GPM method, namely (1) run-to-failure data from multiple units are available and (2) the history of each training unit ends at a soft failure, but may begin at some random level of initial degradation. However, this similarity-based method suffers the same deficiency that all similarity-based models suffer; it is only applicable within the range of data used for training. The proposed GPM/Bayes method will trend toward the training data when the Bayesian information is dominant, early in equipment degradation or when data are very noisy, but as more data become available, the method will accommodate degradation paths outside those seen in training. Additionally, the proposed similarity method requires storage of a large bank of historical data. This may not be a problem for large computer systems, but it can become cumbersome for onboard prognostic algorithms and systems with many fault modes requiring many historical paths. Conversely, the GPM/Bayes method requires storage for only the regression model to be fit and the Bayesian prior information.

The second and third place submissions both focused on recurrent neural networks for prognostic estimation (Heimes, 2008; Peel, 2008). Neural networks require a certain level of expertise and finesse to develop. While they are very powerful modeling tools, neural networks lack the accessibility of the GPM/Bayes method or other regression models. A well-developed neural network may outperform many other prognostic algorithms, but development is not a trivial task. Neural network approaches should not be discounted by any means, but the advantage of the GPM/Bayes method is its relative simplicity.

ACKNOWLEDGEMENT

We would like to acknowledge funding from Ridgetop Group, Inc. under Grant Number NAV-UTK-0207, entitled "Prognostics and Health Management (PHM) for Digital Electronics Using Existing Parameters and Measurands." The information and conclusions presented herein are those of the authors and do not necessarily represent the views or positions of Ridgetop Group.

REFERENCES

Callan, R., B. Larder, and J. Sandiford (2006), "An Integrated Approach to the Development of an Intelligent Prognostic Health Management System," *Proc. Of 2006 IEEE Aerospace Conference*, Big Sky, MT.

Carlin, B.P and T.A. Louis (2000), *Bayes and Empirical Bayes Methods for Data Analysis*, 2nd ed. Boca Raton: Chapman and Hall/CRC.

Chinnam, R.B. (1999), "On-line Reliability Estimation of Individual Components, Using Degradation Signals", *IEEE Transactions on Reliability*, Vol 48, No 4, pp. 403-412.

Garvey, D. and J.W. Hines (2006), "Robust Distance Measures for On-Line Monitoring: Why Use Euclidean?" 7th International Fuzzy Logic and Intelligent Technologies in Nuclear Science (FLINS) Conference on Applied Artificial Intelligence.

Garvey, J., R. Seibert, D. Garvey, and J.W. Hines (2007), "Application of On-Line Monitoring Techniques to Nuclear Plant Data," *Nuclear Engineering and Technology*, Vol. 39 No. 2.

Gelman, A., J. Carlin, H. Stern, and D. Rubin (2004), *Bayesian Data Analysis* 2nd ed. Boca Raton: Chapman and Hall/CRC.

Girish, T., S.W. Lam, J.S.R. Jayaram (2003), "Reliability Prediction Using Degradation Data – A Preliminary Study Using Neural Network-based Approach," *Proc.*

European Safety and Reliability Conference (ESREL 2003), Maastricht, The Netherlands, Jun. 15-18.

Heimes, F.O. (2008), "Recurrent Neural Networks for Remaining Useful Life Estimation," PHM '08, Denver CO, Oct 6-9.

Hines, J.W., D. Garvey, R. Seibert, A. Usynin, and S.A. Arndt, (2008) "Technical Review of On-Line Monitoring Techniques for Performance Assessment, Volume 2: Theoretical Issues," NUREG/CR-6895 Vol 2.

Hines, J.W., J. Garvey, J. Preston, and A. Usynin (2007), "Empirical Methods for Process and Equipment Prognostics," *Reliability and Maintainability Symposium* RAMS.

Koppen, M. (2004), "No-Free-Lunch Theorems and the Diversity of Algorithms", *Congress on Evolutionary Computation*, 1, pp 235 – 241.

Kothamasu, R., S.H. Huang, and W.H. VerDuin (2006), "System Health Monitoring and Prognostics – A Review of Current Paradigms and Practices," *International Journal of Advanced Manufacturing Technology* **28**: 1012 – 1024.

Lindely, D.V. and A.F. Smith (1972), "Bayes Estimates for Linear Models," *Journal of the Royal Statistical Society (B)*, Vol 34, No 1, pp. 1-41.

Lu, C.J. and W.Q. Meeker (1993), "Using Degradation Measures to Estimate a Time-to-Failure Distribution," *Technometrics*, Vol 35, No 2, pp. 161-174.

Meeker, W.Q., L.A. Escobar, and C.J. Lu (1998), "Accelerated degradation tests: modeling and analysis," *Technometrics*, vol. 40 **(2)** pp. 89-99.

Oja, M., J.K. Line, G. Krishnan, R.G. Tryon (2007), "Electronic Prognostics with Analytical Models using Existing Measurands," 61st Conference of the Society for Machinery Failure Prevention Technology (MFPT) 2007, Virginia Beach, April 17-19.

Pecht, M. and A. Dasgupta (1995), "Physics of Failure: An Approach to Reliable Product Development," *Journal of the Institute of Environmental Sciences* 38 (5): 30 – 34.

Peel, L. (2008), "Data Driven Prognostics using a Kalman Filter Ensemble of Neural Network Models," PHM '08, Denver CO, Oct 6-9.

Robinson, M.E. and M.T. Crowder (2000), "Bayesian Methods for a Growth-Curve Degradation Model with Repeated Measures," *Lifetime Data Analysis*, Vol 6, pp. 357-374.

Saxena, A., K. Goebel, D. Simon, N. Eklund (2008), "Prognostics Challenge Competition Summary: Damage Propagation Modeling for Aircraft

Engine Run-to-Failure Simulation," PHM '08, Denver CO, Oct 6-9.

Upadhyaya, B.R., M. Naghedolfeizi, and B. Raychaudhuri (1994), "Residual Life Estimation of Plant Components," *P/PM Technology*, June, pp. 22-29.

Valentin, R., M. Osterman, B. Newman (2003), "Remaining Life Assessment of Aging Electronics in Avionic Applications," 2003 Proceedings of the Annual Reliability and Maintainability Symposium (RAMS): 313 – 318.

Wald, A. (1945), "Sequential Tests of Statistical Hypotheses," *The Annals of Mathematical Statistics* June 1945, vol. 16 **(2)** pp. 117-186.

Wang, P. and D.W. Coit (2004), "Reliability Prediction based on Degradation Modeling for Systems with Multiple Degradation Measures," *Proc. of the 2004 Reliability and Maintainability Symposium*, pp. 302-307.

Wang,T. J. Yu, D. Seigel, and J. Lee (2008), "A Similarity-Based Prognostics Approach for Remaining Useful Life Estimation of Engineered Systems," PHM '08, Denver CO, Oct 6-9.

Dr. Jamie B. Coble attended the University of Tennessee, Knoxville where she graduated Summa Cum Laude with a Bachelor of Science degree in both Nuclear Engineering and Mathematics and a minor in Engineering Communication and Performance in May, 2005. She also completed the University Honors Program with a Senior Project titled, "Investigating Neutron Spectra Changes in Deep Penetration Shielding Analyses." In January, 2005, she began work with Dr. Mario Fontana investigating the effects of long-term station blackouts on boiling water reactors, focusing particularly on the radionuclide release to the environment. In October, 2005, she began research with Dr. J. Wesley Hines investigating the effects of poor model construction on auto-associative model architectures for on-line monitoring systems. This work was incorporated into the final volume of a NUREG series for the U.S. Nuclear Regulatory Commission (NRC). She received an MS in Nuclear Engineering for this work in August, 2006. She was the first graduate of the Reliability and Maintenance Engineering MS program in August, 2009. She completed her Ph.D. in Nuclear Engineering in May, 2010 with work focusing on automated methods for identifying appropriate prognostic parameters for use in individual-based prognosis. Her current research is in the area of empirical methods for system monitoring, fault detection and isolation, and prognostics. She is a member of the engineering honors society Tau Beta Pi, the national honors society Omicron Delta Kappa, the Institute for Electrical and Electronics Engineers Reliability Society, and the American Nuclear Society.

Dr. J. Wesley Hines is a Professor of Nuclear Engineering at the University of Tennessee and is the director of the Reliability and Maintainability Engineering Education program. He received the BS degree in Electrical Engineering from Ohio University in 1985, and then was a nuclear qualified submarine officer in the Navy. He received both an MBA and an MS in Nuclear Engineering from The Ohio State University in 1992, and a Ph.D. in Nuclear Engineering from The Ohio State University in 1994. Dr. Hines teaches and conducts research in artificial intelligence and advanced statistical techniques applied to process diagnostics, condition based maintenance, and prognostics. Much of his research program involves the development of algorithms and methods to monitor high value equipment, detect abnormalities, and predict time to failure. He has authored over 250 papers and has several patents in the area of advanced process monitoring and prognostics techniques. He is a director of the Prognostics and Health Management Society, and a member of the American Nuclear Society, American Society of Engineering Education.

INTERNATIONAL JOURNAL OF PROGNOSTICS AND HEALTH MANAGEMENT, VOL.2 (2011)

A Model-Based Prognostics Approach Applied to Pneumatic Valves*

Matthew J. Daigle [1] and Kai Goebel [2]

[1] *University of California, Santa Cruz, NASA Ames Research Center, Moffett Field, CA, 94035, USA*
matthew.j.daigle@nasa.gov
[2] *NASA Ames Research Center, Moffett Field, CA, 94035, USA*
kai.goebel@nasa.gov

ABSTRACT

Within the area of systems health management, the task of prognostics centers on predicting when components will fail. Model-based prognostics exploits domain knowledge of the system, its components, and how they fail by casting the underlying physical phenomena in a physics-based model that is derived from first principles. Uncertainty cannot be avoided in prediction, therefore, algorithms are employed that help in managing these uncertainties. The particle filtering algorithm has become a popular choice for model-based prognostics due to its wide applicability, ease of implementation, and support for uncertainty management. We develop a general model-based prognostics methodology within a robust probabilistic framework using particle filters. As a case study, we consider a pneumatic valve from the Space Shuttle cryogenic refueling system. We develop a detailed physics-based model of the pneumatic valve, and perform comprehensive simulation experiments to illustrate our prognostics approach and evaluate its effectiveness and robustness. The approach is demonstrated using historical pneumatic valve data from the refueling system.

1 INTRODUCTION

Prognostics is concerned with determining the health of system components and making *end of life* (EOL) and *remaining useful life* (RUL) predictions. It is a key enabling technology for condition-based maintenance, and serves to increase system availability, reliability, and safety by enabling timely maintenance decisions to be made. As with diagnos-

tics, prognostics methods are typically categorized as either model-based or data-driven. Data-driven approaches, rather than taking advantage of system and domain knowledge, instead utilize large amounts of run-to-failure data that are used to train machine learning algorithms to identify trends and determine EOL and RUL (Schwabacher, 2005). However, such data is often difficult to acquire. In contrast, model-based approaches exploit domain knowledge of the system, its components, and how they fail, in the form of models, in order to provide EOL and RUL predictions (Roemer, Byington, Kacprzynski, & Vachtsevanos, 2005; Byington, Watson, Edwards, & Stoelting, 2004; Saha & Goebel, 2009; Daigle & Goebel, 2010b; Luo, Pattipati, Qiao, & Chigusa, 2008). The underlying physical phenomena are captured in a physics-based model that is derived from first principles, therefore, model-based approaches can provide EOL and RUL estimates that are much more accurate and precise than data-driven approaches, if the models are correct.

In this paper, we develop a general model-based prognostics framework using particle filters, based on preliminary work presented in (Daigle & Goebel, 2009, 2010b). Particle filters are nonlinear state observers that approximate the posterior state distribution as a set of discrete, weighted samples. They have become a popular methodology in the context of prognostics, where they are used for joint state-parameter estimation, e.g., (Saha & Goebel, 2009; Orchard, 2007; Daigle & Goebel, 2009, 2010b). Although suboptimal, the advantage of particle filters is that they can be applied to systems which may be nonlinear and have non-Gaussian noise terms, where optimal solutions are unavailable or intractable. Further, because they are based on probability distributions, they help in managing the uncertainty that may arise from a number of sources in prognostics.

As a case study, we consider a pneumatic valve from the Space Shuttle cryogenic refueling system. We construct a detailed physics-based model of a pneumatic valve that in-

---

*The funding for this work was provided by the NASA Fault Detection, Isolation, and Recovery (FDIR) project under the Exploration Technology and Development Program (ETDP) of the Exploration Systems Mission Directorate (ESMD).

cludes models of different damage mechanisms. We run a comprehensive set of prognostics experiments in simulation to demonstrate the approach and establish that prognostics may be performed for pneumatic valves using only discrete position sensors. Further, we demonstrate the approach using historical pneumatic valve data from the refueling system.

The paper is organized as follows. Section 2 formally defines the prognostics problem and describes the computational architecture. Section 3 presents the modeling methodology and develops the model of the pneumatic valve. Section 4 discusses the damage estimation approach using particle filters, and Section 5 provides the prediction algorithm. Section 6 presents simulation results and the demonstration of the approach on real data. Section 7 discusses related work. Section 8 concludes the paper.

2 PROGNOSTICS APPROACH

The problem of prognostics is predicting the EOL and/or the RUL of a component. In this section, we first formally define the problem of model-based prognostics. We then describe a general model-based architecture within which a prognostics solution may be implemented.

2.1 Problem Formulation

We assume the system may be described by

$$\dot{\mathbf{x}}(t) = \mathbf{f}(t, \mathbf{x}(t), \boldsymbol{\theta}(t), \mathbf{u}(t), \mathbf{v}(t))$$
$$\mathbf{y}(t) = \mathbf{h}(t, \mathbf{x}(t), \boldsymbol{\theta}(t), \mathbf{u}(t), \mathbf{n}(t))$$

where $t \in \mathbb{R}$ is the continuous time variable, $\mathbf{x}(t) \in \mathbb{R}^{n_x}$ is the state vector, $\boldsymbol{\theta}(t) \in \mathbb{R}^{n_\theta}$ is the parameter vector, $\mathbf{u}(t) \in \mathbb{R}^{n_u}$ is the input vector, $\mathbf{v}(t) \in \mathbb{R}^{n_v}$ is the process noise vector, $\mathbf{f}$ is the state equation, $\mathbf{y}(t) \in \mathbb{R}^{n_y}$ is the output vector, $\mathbf{n}(t) \in \mathbb{R}^{n_n}$ is the measurement noise vector, and $\mathbf{h}$ is the output equation. This representation considers a general nonlinear model with no restrictions on the functional forms of $\mathbf{f}$ or $\mathbf{h}$. Further, the noise terms may be coupled in a nonlinear way with the states and parameters. The parameters $\boldsymbol{\theta}(t)$ evolve in an unknown way. In practice, they are typically considered to be constant, but may in fact be time-varying.

Our goal is to predict EOL at a given time point t_P using the discrete sequence of observations up to time t_P, denoted as $\mathbf{y}_{0:t_P}$. EOL is defined as the time point at which the component no longer meets one of a set of functional requirements (e.g., a valve does not open in the required amount of time). These requirements may be expressed using a threshold, beyond which we say the component has failed. In general, we may express this threshold as a function of the system state and parameters, $T_{EOL}(\mathbf{x}(t), \boldsymbol{\theta}(t))$, which determines whether the system has failed, where $T_{EOL}(\mathbf{x}(t), \boldsymbol{\theta}(t)) = 1$ if a requirement is violated, and 0 otherwise. Using this function, we can formally define EOL with

$$EOL(t_P) \triangleq \inf\{t \in \mathbb{R} : t \geq t_P \wedge T_{EOL}(\mathbf{x}(t), \boldsymbol{\theta}(t)) = 1\},$$

and RUL with

$$RUL(t_P) \triangleq EOL(t_P) - t_P.$$

In practice, many sources of uncertainty exist that unfold into the prediction. Uncertainty in the initial state, uncertainty in the model, process noise (i.e., $\mathbf{v}(t)$), and sensor noise (i.e., $\mathbf{n}(t)$) result in an uncertain estimate of $(\mathbf{x}(t), \boldsymbol{\theta}(t))$. In predicting from this uncertain state, modeling errors, process noise, and uncertainty in the future inputs of the system further add to the overall uncertainty. At best, then, we can only compute a probability distribution of the EOL or RUL, rather than a single prediction point. The goal, then, is to compute, at time t_P, $p(EOL(t_p)|\mathbf{y}_{0:t_P})$ and/or $p(RUL(t_P)|\mathbf{y}_{0:t_P})$.

2.2 Prognostics Architecture

We adopt a model-based approach, wherein we develop detailed physics-based models of components and systems that include descriptions of how faults evolve in time. These models depend on unknown and possibly time-varying parameters. Therefore, our solution to the prognostics problem takes the perspective of joint state-parameter estimation. In discrete time k, we estimate $\mathbf{x}_k$ and $\boldsymbol{\theta}_k$, and use these estimates to predict EOL and RUL at desired time points.

We employ the prognostics architecture in Fig. 1. The system is provided with inputs $\mathbf{u}_k$ and provides measured outputs $\mathbf{y}_k$. Prognostics may begin at $k = 0$, with the damage estimation module determining estimates of the states and unknown parameters, represented as a probability distribution $p(\mathbf{x}_k, \boldsymbol{\theta}_k|\mathbf{y}_{0:k})$. The prediction module uses the joint state-parameter distribution, along with hypothesized future inputs, to compute EOL and RUL as probability distributions $p(EOL_{k_P}|\mathbf{y}_{0:k_P})$ and $p(RUL_{k_P}|\mathbf{y}_{0:k_P})$ at given prediction times k_P. In parallel, a fault detection, isolation, and identification (FDII) module may be used to determine which damage mechanisms are active, represented as a fault set $\mathbf{F}$. The damage estimation module may then use this result to limit the space of parameters that must be estimated. Alternatively, prognostics may begin only when diagnostics has completed. In this paper, we assume prognostics is not aided by FDII.

3 MODELING METHODOLOGY

To implement a model-based prognostics approach, detailed physics-based models of the component are necessary. Such models must not only describe the nominal behavior of the component, but also the faulty behavior. Further, they must describe how faults, or damage, progress in time. It is with these models that prediction may be performed.

To construct models in this paradigm, the first step is to develop a nominal model based on a first principles, physical understanding of the system. The next step is to identify the variables or parameters that are representative of faults, which we call damage variables, $\mathbf{d}(t)$, e.g., a friction parameter, or the size of a leak. Since these variables change dynamically,

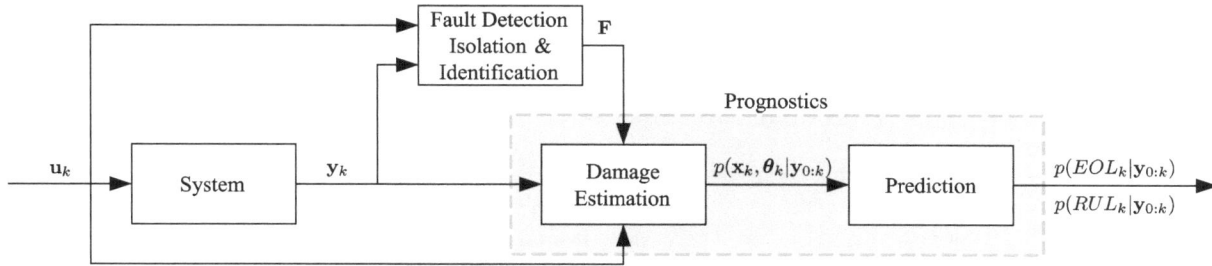

Figure 1. Prognostics architecture

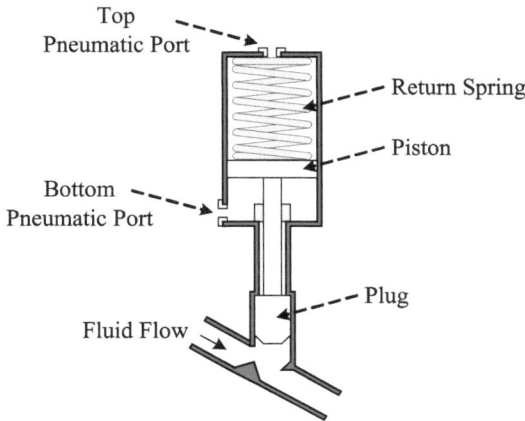

Figure 2. Pneumatic valve

they form part of the state vector, i.e., $\mathbf{d}(t) \subseteq \mathbf{x}(t)$. The final step is to develop models for how these variables change in time, i.e., how the damage progresses or evolves, which we call *damage progression functions*. These models are typically dependent on other states of the system, and, therefore, dependent on the system inputs. They augment the state equation $\mathbf{f}$. These functions are parameterized by unknown parameters, which we call *wear parameters*, $\mathbf{w}(t) \subseteq \boldsymbol{\theta}(t)$.

In the remainder of this section, we apply this modeling framework to a pneumatic valve, which serves as the case study for this paper.

3.1 Pneumatic Valve Modeling

Pneumatic valves are gas-actuated valves used in many domains. A normally-closed valve with a linear cylinder actuator is illustrated in Fig. 2. The valve is opened by filling the chamber below the piston with gas up to the supply pressure, and evacuating the chamber above the piston down to atmospheric pressure. The valve is closed by filling the chamber above the piston, and evacuating the chamber below the piston. The return spring ensures that when pressure is lost, the valve will close due to the force exerted by the return spring, hence it is a normally-closed valve.

We develop a physics model of the valve based on mass and energy balances. The system state for the nominal model

includes the position of the valve, $x(t)$, the velocity of the valve, $v(t)$, the mass of the gas in the volume above the piston, $m_t(t)$, and the mass of the gas in the volume below the piston, $m_b(t)$:

$$\mathbf{x}(t) = \begin{bmatrix} x(t) & v(t) & m_t(t) & m_b(t) \end{bmatrix}^T.$$

The position is defined as $x = 0$ when the valve is fully closed. The stroke length of the valve is denoted by L_s; when the valve is fully open its position $x = L_s$.

The derivatives of the states are described by

$$\dot{\mathbf{x}}(t) = \begin{bmatrix} v(t) & a(t) & f_t(t) & f_b(t) \end{bmatrix}^T,$$

where $a(t)$ is the valve acceleration, and $f_t(t)$ and $f_b(t)$ are the mass flows going into the top and bottom pneumatic ports, respectively.

The inputs are considered to be

$$\mathbf{u}(t) = \begin{bmatrix} p_l(t) & p_r(t) & u_t(t) & u_b(t) \end{bmatrix}^T,$$

where $p_l(t)$ and $p_r(t)$ are the fluid pressures on the left and right side of the plug, respectively, and $u_t(t)$ and $u_b(t)$ are the input pressures to the top and bottom pneumatic ports. These pressures will alternate between the supply pressure and atmospheric pressure depending on the commanded valve position.

The acceleration is defined by the combined mass of the piston and plug, m, and the sum of forces acting on the valve, which includes the forces from the pneumatic gas, $(p_b(t) - p_t(t))A_p$, where $p_b(t)$ and $p_t(t)$ are the gas pressures on the bottom and the top of the piston, respectively, and A_p is the surface area of the piston; the forces from the fluid flowing through the valve, $(p_r(t) - p_l(t))A_v$, where A_v is the area of the valve contacting the fluid; the weight of the moving parts of the valve, $-mg$, where g is the acceleration due to gravity; the spring force, $-k(x(t) - x_o)$, where k is the spring constant and x_o is the amount of spring compression when the valve is closed; friction, $-rv(t)$, where r is the coefficient of kinetic friction, and the contact forces $F_c(t)$ at the boundaries of the valve motion,

$$F_c(t) = \begin{cases} k_c(-x), & \text{if } x < 0, \\ 0, & \text{if } 0 \leq x \leq L_s, \\ -k_c(x - L_s), & \text{if } x > L_s, \end{cases}$$

where k_c is the (large) spring constant associated with the flexible seals. Overall, the acceleration term is defined by

$$a(t) = \frac{1}{m}\Big[(p_b(t) - p_t(t))A_p + (p_r(t) - p_l(t))A_v$$
$$- mg - k(x(t) - x_o) - rv(t) + F_c(t)\Big].$$

The pressures $p_t(t)$ and $p_b(t)$ are calculated as:

$$p_t(t) = \frac{m_t(t)R_gT}{V_{t_0} + A_p(L_s - x(t))}$$
$$p_b(t) = \frac{m_b(t)R_gT}{V_{b_0} + A_px(t)},$$

where we assume an isothermal process in which the (ideal) gas temperature is constant at T, R_g is the gas constant for the pneumatic gas, and V_{t_0} and V_{b_0} are the minimum gas volumes for the gas chambers above and below the piston, respectively. The gas flows are given by:

$$f_t(t) = f_g(p_t(t), u_t(t))$$
$$f_b(t) = f_g(p_b(t), u_b(t)),$$

where f_g defines gas flow through an orifice for choked and non-choked flow conditions (Perry & Green, 2007). Non-choked flow for $p_1 \geq p_2$ is given by $f_{g,nc}(p_1, p_2) =$

$$C_sA_sp_1\sqrt{\frac{\gamma}{ZR_gT}\left(\frac{2}{\gamma - 1}\right)\left(\left(\frac{p_2}{p_1}\right)^{\frac{2}{\gamma}} - \left(\frac{p_2}{p_1}\right)^{\frac{\gamma+1}{\gamma}}\right)},$$

where γ is the ratio of specific heats, Z is the gas compressibility factor, C_s is the flow coefficient, and A_s is the orifice area. Choked flow for $p_1 \geq p_2$ is given by

$$f_{g,c}(p_1, p_2) = C_sA_sp_1\sqrt{\frac{\gamma}{ZR_gT}\left(\frac{2}{\gamma + 1}\right)^{\frac{\gamma+1}{\gamma-1}}}.$$

Choked flow occurs when the upstream to downstream pressure ratio exceeds $\left(\frac{\gamma+1}{2}\right)^{\gamma/(\gamma-1)}$. The overall gas flow equation is then given by

$$f_g(p_1, p_2) = \begin{cases} f_{g,nc}(p_1, p_2) & \text{if } p_1 \geq p_2 \\ & \text{and } \frac{p_1}{p_2} < \left(\frac{\gamma+1}{2}\right)^{\frac{\gamma}{(\gamma-1)}}, \\ f_{g,c}(p_1, p_2) & \text{if } p_1 \geq p_2 \\ & \text{and } \frac{p_1}{p_2} \geq \left(\frac{\gamma+1}{2}\right)^{\frac{\gamma}{(\gamma-1)}}, \\ -f_{g,nc}(p_2, p_1) & \text{if } p_2 > p_1 \\ & \text{and } \frac{p_2}{p_1} < \left(\frac{\gamma+1}{2}\right)^{\frac{\gamma}{(\gamma-1)}}, \\ -f_{g,c}(p_2, p_1) & \text{if } p_2 > p_1 \\ & \text{and } \frac{p_2}{p_1} \geq \left(\frac{\gamma+1}{2}\right)^{\frac{\gamma}{(\gamma-1)}}, \end{cases}$$

We select our complete measurement vector as

$$\mathbf{y}(t) = \begin{bmatrix} x(t) & p_t(t) & p_b(t) & f_v(t) & open(t) & closed(t) \end{bmatrix}^T.$$

Here, f_v is the fluid flow through the valve, given by

$$f_v(t) = \frac{x(t)}{L_s}C_vA_v\sqrt{\frac{2}{\rho}|p_{fl} - p_{fr}|}\text{sign}(p_{fl} - p_{fr}),$$

where C_v is the (dimensionless) flow coefficient of the valve, ρ is the liquid density, and we assume a linear flow characteristic for the valve. The $open(t)$ and $closed(t)$ signals are from discrete sensors that output 1 if the valve is in the fully open or fully closed state:

$$open(t) = \begin{cases} 1, & \text{if } x(t) \geq L_s \\ 0, & \text{otherwise} \end{cases}$$
$$closed(t) = \begin{cases} 1, & \text{if } x(t) \leq 0 \\ 0, & \text{otherwise.} \end{cases}$$

Fig. 3 shows a nominal valve cycle. The valve is commanded to open at 0 s. The top pneumatic port opens to atmosphere and the bottom port opens to the supply pressure (approximately 5.3 MPa, or 750 psig). When the force on the underside of the piston is large enough to overcome the return spring, friction, and the gas force on the top of the piston, the valve begins to move upward as the pneumatic gas continues to flow into and out of the valve actuator. At about 8 s, the valve is completely open. The valve is commanded to close at 15 s. The bottom pneumatic port opens to atmosphere and the top port opens to the supply pressure. When the force balance becomes negative, the valve starts to move downward, and completely closes at around 20 s. The valve closes faster than it opens due to the return spring.

3.2 Damage Modeling

With the nominal model defined, we may now determine which damage mechanisms are relevant, and what model parameters change as a result of the damage. From valve documentation and historical maintenance records, we have identified a set of faults that, although not exhaustive, contains the most important and most often observed faults that affect valve functionality and lead to EOL. The set of faults includes friction damage, spring damage, internal valve leaks, and external valve leaks. The functional requirements on the valve that define T_{EOL} are that it opens and closes within given timing limits, and that it fully closes upon loss of actuating pressure. In this example, we use 15 s for both the opening and closing time limits.

One damage mechanism present in valves is sliding wear. The equation for sliding wear takes on the following form (Hutchings, 1992):

$$\dot{V}(t) = w|F(t)v(t)|,$$

where $V(t)$ is the wear volume, w is the wear coefficient (which depends on material properties such as hardness), $F(t)$ is the sliding force, and $v(t)$ is the sliding velocity. Friction will increase linearly with sliding wear, because the

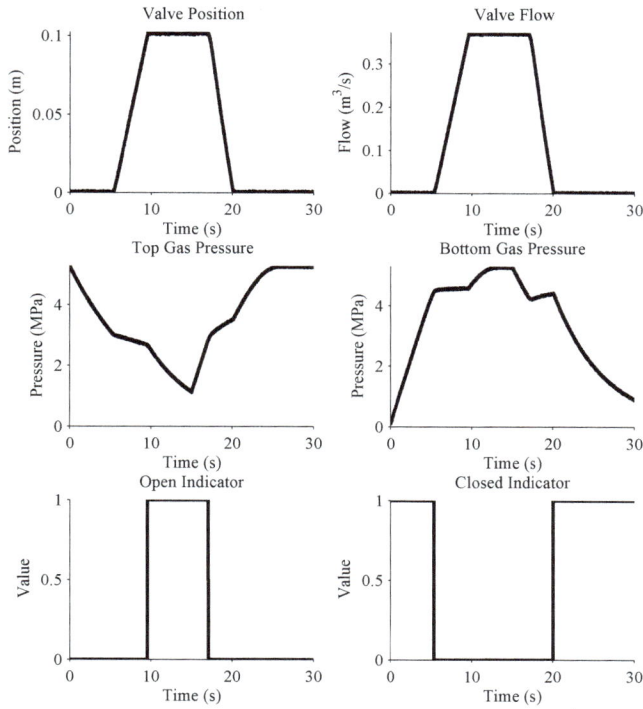

Figure 3. Nominal valve operation

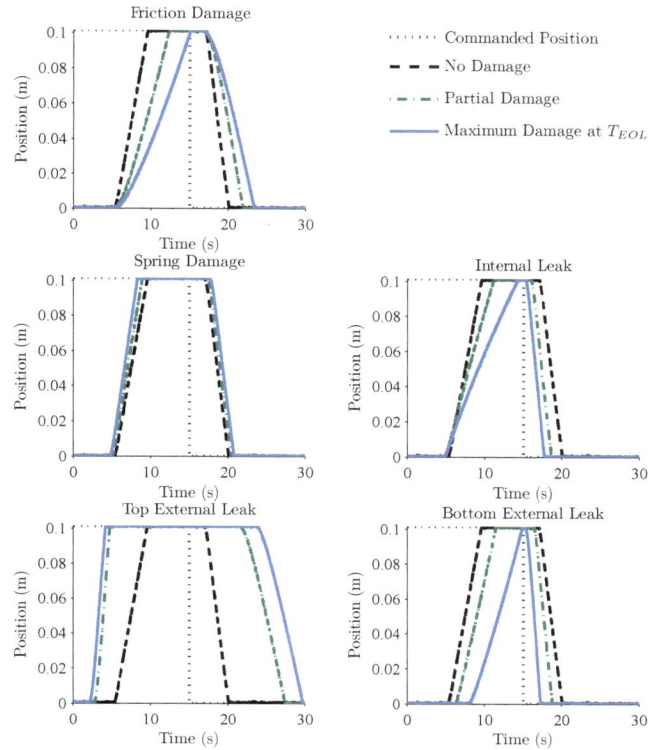

Figure 4. Valve operation with damage

contact area between the sliding bodies becomes greater as surface asperities wear down (Hutchings, 1992). Lubrication between the sliding bodies can also degrade over time. We therefore characterize friction damage as change in the friction coefficient, and model the damage progression in a form similar to sliding wear:

$$\dot{r}(t) = w_r |F_f(t)v(t)|$$

where w_r is the wear coefficient, and $F_f(t)$ is the friction force defined in the previous subsection. The friction coefficient only grows when the valve is moving, as only then can sliding wear occur. Therefore, the friction coefficient evolves in a step-wise fashion, with damage only growing during the opening and closing motions. As the friction coefficient increases, the friction force increases, further increasing the rate at which the friction coefficient grows. This results in a damage progression that is similar to an exponential when viewed at large time scales.

Fig. 4 shows the effect of an increase in friction on the valve cycle. We define r^+ as the maximum value of the friction parameter at which the valve still behaves within the timing limits. Above this value, the friction force becomes large enough that the valve cannot open within the 15 s limit, as shown in Fig. 4. So, $T_{EOL}(\mathbf{x}(t), \boldsymbol{\theta}(t)) = 1$ if $r(t) > r^+$.

We assume a similar equation form for spring damage:

$$\dot{k}(t) = -w_k |F_s(t)v(t)|,$$

where w_k is the spring wear coefficient and $F_s(t)$ is the spring

force. The more the spring is used, the weaker it becomes, characterized by the change in the spring constant. Like with friction damage, the spring constant decreases only when the valve moves. As the spring becomes damaged, the spring force will decrease, and so the rate at which spring damage occurs will also decrease.

Fig. 4 shows the effect of a decrease in the spring constant on the valve cycle. In normal operation, without the spring tending the valve to close, the valve will open faster and close slower. The spring can actually fail completely, and the valve would still open and close in time, however, the spring must be strong enough to close the valve against system pressure when the actuating pressure is lost. So, it is the loss of this function leads to T_{EOL} for spring damage, and we define k^- as the smallest value of k at which the valve will still fully close upon loss of supply pressure. So, $T_{EOL}(\mathbf{x}(t), \boldsymbol{\theta}(t)) = 1$ also if $k(t) < k^-$.

An internal leak in the valve can appear at the seal surrounding the piston as a result of sliding wear. The pneumatic gas is then able to flow between the volumes above and below the piston, decreasing the response time of the valve. We parameterize this leak by its equivalent orifice area, $A_i(t)$, described by

$$\dot{A}_i(t) = w_i |F_f(t)v(t)|,$$

where w_i is the wear coefficient. The mass flow at the leak,

$f_i(t)$, is computed using the gas flow equation described earlier:

$$f_i(t) = f_g(p_t(t), p_b(t)),$$

where positive flow denotes flow from the top volume to the bottom volume. The term is subtracted from the $f_t(t)$ equation and added to the $f_b(t)$ equation.

As sliding wear occurs, the leak size increases. As with the friction and spring damage, the internal leak only grows when the valve is moving. Over large time scales, the internal leak appears to grow linearly with the valve cycles, since the friction force does not change much as the leak size grows.

The presence of an internal leak makes it more difficult to actuate the valve, because it causes gas to flow into the lower pressure volume that is being evacuated and out of the higher pressure volume that is being filled. We define A_i^+ as the maximum internal leak area at which the valve opens and closes within the functional limits. So, $T_{EOL}(\mathbf{x}(t), \boldsymbol{\theta}(t)) = 1$ also if $A_i(t) > A_i^+$. Fig. 4 shows the effect of an internal leak on the valve cycle.

External leaks may also form, most likely at the actuator connections to the pneumatic gas supply, due to corrosion and other environmental factors. Without knowledge of how the leak size progresses, we assume the growth of the area of the leak holes, $A_{e,t}(t)$ and $A_{e,b}(t)$, is linear:

$$\dot{A}_{e,t}(t) = w_{e,t}$$
$$\dot{A}_{e,b}(t) = w_{e,b},$$

where the t and b subscripts denote a leak at the top and bottom pneumatic ports, respectively, and $w_{e,t}$ and $w_{e,b}$ are the wear coefficients. The leak flows are given by

$$f_{e,t}(t) = f_g(p_t(t), p_{atm}(t))$$
$$f_{e,b}(t) = f_g(p_b(t), p_{atm}(t)),$$

where p_{atm} is atmospheric pressure. The $f_{e,t}(t)$ term is subtracted from the $f_t(t)$ equation, and the $f_{e,b}(t)$ term is subtracted from the $f_b(t)$ equation.

Note that this damage progression is independent of the valve inputs. Fig. 4 shows the effects of top and bottom external leaks on the valve cycle. The effect of the formation of a leak at the top pneumatic port is that it becomes easier to open the valve but more difficult to close it. Conversely, the effect of a leak at the bottom pneumatic port is that it becomes more difficult to open but easier to close the valve. We define the maximum leak hole areas at which the valve still opens and closes within the functional limits as $A_{e,t}^+$ and $A_{e,b}^+$. So, $T_{EOL}(\mathbf{x}(t), \boldsymbol{\theta}(t)) = 1$ also if $A_{e,t}(t) > A_{e,t}^+$ or $A_{e,b}(t) > A_{e,b}^+$. Alternatively, maximum allowable leakage rates may define EOL.

So, the damage variables are given by

$$\mathbf{d}(t) = \begin{bmatrix} r(t) & k(t) & A_i(t) & A_{e,t}(t) & A_{e,b}(t) \end{bmatrix}^T,$$

and the complete state vector of the extended valve model becomes

$$\mathbf{x}(t) = \begin{bmatrix} x(t)v(t) \\ m_t(t) \\ m_b(t) \\ r(t) \\ k(t) \\ A_i(t) \\ A_{e,t}(t) \\ A_{e,b}(t) \end{bmatrix}.$$

The wear parameters form the unknown parameter vector, i.e.,

$$\mathbf{w}(t) = \boldsymbol{\theta}(t) = \begin{bmatrix} w_r(t) \\ w_k(t) \\ w_i(t) \\ w_{e,t}(t) \\ w_{e,b}(t) \end{bmatrix}.$$

4 DAMAGE ESTIMATION

In the model-based paradigm, damage estimation reduces to joint state-parameter estimation, i.e., computation of $p(\mathbf{x}_k, \boldsymbol{\theta}_k | \mathbf{y}_{0:k})$. The typical approach to state estimation is to use a state observer, which helps to compensate for the effects of process and sensor noise on the state estimate. Joint state-parameter estimation is traditionally performed using a state observer by augmenting the state vector with the parameter vector. In this way, both states and parameters are simultaneously estimated.

For linear systems with additive Gaussian noise terms, the Kalman filter is appropriate. For nonlinear systems with additive Gaussian noise terms, the extended Kalman filter or unscented Kalman filter (Julier & Uhlmann, 1997) may be used. However, for nonlinear systems with non-Gaussian noise terms, *particle filters* are best suited, and offer approximate (suboptimal) solutions to the state estimation problem for such systems where optimal solutions are unavailable or intractable (Arulampalam, Maskell, Gordon, & Clapp, 2002; Cappe, Godsill, & Moulines, 2007). In particle filters, the state distribution is approximated by a set of discrete weighted samples, called *particles*. As the number of particles is increased, accuracy increases and the optimal solution is approached. In addition, particle filters are straightforward to implement, and computational complexity may be controlled by increasing or decreasing the number of particles, with respect to the desired estimation performance. Further, the discrete position sensors cannot be directly handled by other filtering algorithms. For these reasons, we select particle filters for our model-based prognostics framework.

With particle filters, the particle approximation to the state distribution is given by

$$\{(\mathbf{x}_k^i, \boldsymbol{\theta}_k^i), w_k^i\}_{i=1}^N,$$

where N denotes the number of particles, and for particle i, $\mathbf{x}_k^i$ denotes the state estimates, $\boldsymbol{\theta}_k^i$ denotes the parameter es-

INTERNATIONAL JOURNAL OF PROGNOSTICS AND HEALTH MANAGEMENT, VOL.2 (2011)

Algorithm 1 SIR Filter

Inputs: $\{(\mathbf{x}_{k-1}^i, \boldsymbol{\theta}_{k-1}^i), w_{k-1}^i\}_{i=1}^N, \mathbf{u}_{k-1:k}, \mathbf{y}_k$
Outputs: $\{(\mathbf{x}_k^i, \boldsymbol{\theta}_k^i), w_k^i\}_{i=1}^N$
for $i = 1$ **to** N **do**
    $\boldsymbol{\theta}_k^i \sim p(\boldsymbol{\theta}_k | \boldsymbol{\theta}_{k-1}^i)$
    $\mathbf{x}_k^i \sim p(\mathbf{x}_k | \mathbf{x}_{k-1}^i, \boldsymbol{\theta}_{k-1}^i, \mathbf{u}_{k-1})$
    $w_k^i \leftarrow p(\mathbf{y}_k | \mathbf{x}_k^i, \boldsymbol{\theta}_k^i, \mathbf{u}_k)$
end for
$W \leftarrow \sum_{i=1}^N w_k^i$
for $i = 1$ **to** N **do**
    $w_k^i \leftarrow w_k^i / W$
end for
$\{(\mathbf{x}_k^i, \boldsymbol{\theta}_k^i), w_k^i\}_{i=1}^N \leftarrow$ Resample$(\{(\mathbf{x}_k^i, \boldsymbol{\theta}_k^i), w_k^i\}_{i=1}^N)$

Algorithm 2 Systematic Resampling

Inputs: $\{(\mathbf{x}_k^i, \boldsymbol{\theta}_k^i), w_k^i\}_{i=1}^N$
Outputs: $\{(\mathbf{x}_k^j, \boldsymbol{\theta}_k^j), w_k^j\}_{j=1}^N$
$\mathbf{c}_1 \leftarrow 0$
for $i = 2$ **to** N **do**
    $\mathbf{c}_i \leftarrow \mathbf{c}_{i-1} + w_k^i$
end for
$i \leftarrow 1$
$u_1 \sim \mathcal{U}(0, 1/N)$
for $j = 1$ **to** N **do**
    $u \leftarrow u_1 + (j+1)/N$
    while $u > \mathbf{c}_i$ **do**
        $i \leftarrow i + 1$
    end while
    $\mathbf{x}_k^j \leftarrow \mathbf{x}_k^i$
    $\boldsymbol{\theta}_k^j \leftarrow \boldsymbol{\theta}_k^i$
    $w_k^j = 1/N$
end for

---

timates, and w_k^i denotes the weight. The posterior density is approximated by

$$p(\mathbf{x}_k, \boldsymbol{\theta}_k | \mathbf{y}_{0:k}) \approx \sum_{i=1}^N w_k^i \delta_{(\mathbf{x}_k^i, \boldsymbol{\theta}_k^i)}(d\mathbf{x}_k d\boldsymbol{\theta}_k),$$

where $\delta_{(\mathbf{x}_k^i, \boldsymbol{\theta}_k^i)}(d\mathbf{x}_k d\boldsymbol{\theta}_k)$ denotes the Dirac delta function located at $(\mathbf{x}_k^i, \boldsymbol{\theta}_k^i)$.

We employ the sampling importance resampling (SIR) particle filter, and implement the resampling step using systematic resampling (Kitagawa, 1996). The pseudocode for a single step of the SIR filter is shown as Algorithm 1. Each particle i is propagated forward to time k by first sampling new parameter values. Here, the parameters $\boldsymbol{\theta}_k$ evolve by some unknown random process that is independent of the state $\mathbf{x}_k$. To perform parameter estimation within a particle filter framework, however, we need to assign some type of evolution to the parameters. The typical solution is to use a random walk, i.e., for parameter θ, $\theta_k = \theta_{k-1} + \xi_{k-1}$, where ξ_{k-1} is typically Gaussian noise. After sampling parameter values from the selected distribution, new states are sampled by applying the state equation $\mathbf{f}$ to $(\mathbf{x}_k^i, \boldsymbol{\theta}_k^i)$ with process noise $\mathbf{v}(t)$ sampled from its assumed distribution. The particle weight is assigned using $\mathbf{y}_k$. Specifically, the output equation $\mathbf{h}$ is applied to $(\mathbf{x}_{k+1}^i, \boldsymbol{\theta}_{k+1}^i)$ and the likelihood of the corresponding output is computed using the assumed probability density function of the sensor noise. The weights are then normalized, followed by the resampling step.

Pseudocode for the resampling step is given as Algorithm 2. First, the cumulative distribution function (CDF) of the weights is computed as $\mathbf{c}$. A starting point is drawn from the uniform distribution. The algorithm then moves along the CDF and a new particle is resampled from the old particle set. The weights of the resampled particles are all assigned to be the same. The resampling step is necessary to avoid *degeneracy*, in which all but a few particles have negligible weight. If this occurs, then the state distribution is very poorly represented, and computation is wasted on particles that do not contribute to the approximation. Resampling based on the

CDF avoids this problem by multiplying particles with high weight and dropping particles with low weight. Here, we resample at each time step, but the frequency of resampling may be reduced by computing the effective number of particles, and only resampling when this number falls below a given threshold (Arulampalam et al., 2002).

During the sampling step, particles are generated with parameter values that will be different from the initial guesses for the unknown parameters. The particles with parameter values closest to the true values should match the outputs better, and therefore be assigned higher weight. Resampling will cause more particles to be generated around these better values. As this process is repeated over time, the particle filter converges to the true values. The selected variance of the random walk noise must be large enough so as to allow convergence in a reasonable amount of time, but small enough such that when convergence is reached, the parameter can be tracked smoothly. Fig. 5 illustrates the effect of the variance value on estimation performance shown in an accuracy-precision plot. The horizontal axis provides estimation accuracy as computed using percent root mean square error (PRMSE) of the unknown wear parameter, $w_{e,b}$, and the vertical axis provides the variability of the wear parameter distribution as computed using relative median absolute deviation (RMAD) (performance metrics are defined in the Appendix). The optimal point is located at the origin. As the random walk variance of $w_{e,b}$ is decreased, both PRMSE and RMAD become smaller, improving in both performance dimensions. As the random walk variance is decreased beyond 5×10^{-20}, however, the wear parameter estimate does not always converge and cannot be tracked. Since the parameter values are unknown, an optimal value cannot be determined a priori. So, selecting an appropriate value can be difficult, but knowledge of the correct order of magnitude of the parameter is helpful. Additionally, correction loop meth-

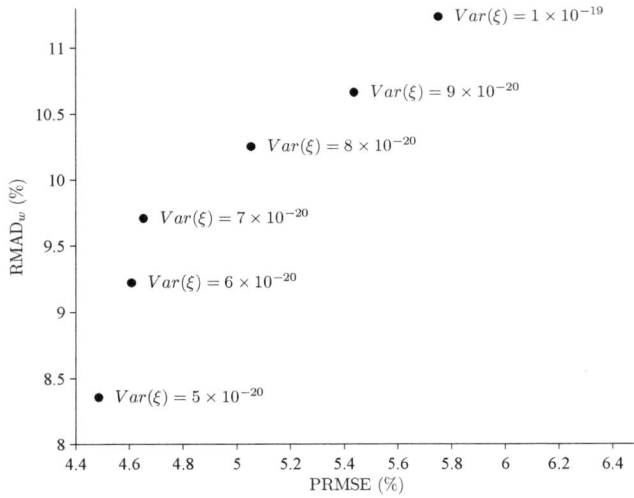

Figure 5. Estimation performance for a bottom external leak, with $N = 500$, $M = \{x, f, p_t, p_b\}$, and varying random walk variance

ods can be used to tune this value online as a function of performance (Orchard, Kacprzynski, Goebel, Saha, & Vachtsevanos, 2008; Saha & Goebel, 2011; Daigle & Goebel, 2011). If the unknown parameters may be assumed to be constant, then other approaches can be employed to improve estimates and offset the increase in covariance contributed by the random walk (Liu & West, 2001; Clapp & Godsill, 1999).

Note that the discrete position sensors (*open* and *closed*), in reality, have no noise, but a certain amount of sensor noise must be assumed for sensors within the particle filter framework. Therefore, within the algorithm, we assume some sensor noise is present for these sensors.

5 PREDICTION

Prediction is initiated at a given time k_P. Using the current state estimate, $p(\mathbf{x}_k, \boldsymbol{\theta}_k|\mathbf{y}_{0:k})$ the goal is to compute $p(EOL_{k_P}|\mathbf{y}_{0:k_P})$ and $p(RUL_{k_P}|\mathbf{y}_{0:k_P})$. The particle filter computes

$$p(\mathbf{x}_{k_P}, \boldsymbol{\theta}_{k_P}|\mathbf{y}_{0:k_P}) \approx \sum_{i=1}^{N} w_{k_P}^i \delta_{(\mathbf{x}_{k_P}^i, \boldsymbol{\theta}_{k_P}^i)}(d\mathbf{x}_{k_P} d\boldsymbol{\theta}_{k_P}).$$

We can approximate a prediction distribution n steps forward as (Doucet, Godsill, & Andrieu, 2000)

$$p(\mathbf{x}_{k_P+n}, \boldsymbol{\theta}_{k_P+n}|\mathbf{y}_{0:k_P}) \approx$$
$$\sum_{i=1}^{N} w_{k_P}^i \delta_{(\mathbf{x}_{k_P+n}^i, \boldsymbol{\theta}_{k_P+n}^i)}(d\mathbf{x}_{k_P+n} d\boldsymbol{\theta}_{k_P+n}).$$

So, for a particle i propagated n steps forward without new data, we can simply take its weight as $w_{k_P}^i$. Similarly, we can

---

Algorithm 3 EOL Prediction

Inputs: $\{(\mathbf{x}_{k_P}^i, \boldsymbol{\theta}_k^i), w_{k_P}^i\}_{i=1}^{N}$
Outputs: $\{EOL_{k_P}^i, w_{k_P}^i\}_{i=1}^{N}$
for $i = 1$ **to** N **do**
  $k \leftarrow k_P$
  $\mathbf{x}_k^i \leftarrow \mathbf{x}_{k_P}^i$
  $\boldsymbol{\theta}_k^i \leftarrow \boldsymbol{\theta}_{k_P}^i$
  while $T_{EOL}(\mathbf{x}_k^i, \boldsymbol{\theta}_k^i) = 0$ **do**
    Predict $\hat{\mathbf{u}}_k$
    $\boldsymbol{\theta}_{k+1}^i \sim p(\boldsymbol{\theta}_{k+1}|\boldsymbol{\theta}_k^i)$
    $\mathbf{x}_{k+1}^i \sim p(\mathbf{x}_{k+1}|\mathbf{x}_k^i, \boldsymbol{\theta}_k^i, \hat{\mathbf{u}}_k)$
    $k \leftarrow k + 1$
    $\mathbf{x}_k^i \leftarrow \mathbf{x}_{k+1}^i$
    $\boldsymbol{\theta}_k^i \leftarrow \boldsymbol{\theta}_{k+1}^i$
  end while
  $EOL_{k_P}^i \leftarrow k$
end for

---

approximate the EOL as

$$p(EOL_{k_P}|\mathbf{y}_{0:k_P}) \approx \sum_{i=1}^{N} w_{k_P}^i \delta_{EOL_{k_P}^i}(dEOL_{k_P}).$$

To compute EOL, then, we propagate each particle forward to its own EOL and use that particle's weight at k_P for the weight of its EOL prediction.

The pseudocode for the prediction procedure is given as Algorithm 3 (Daigle & Goebel, 2010b). Each particle i is propagated forward until $T_{EOL}(\mathbf{x}_k^i, \boldsymbol{\theta}_k^i)$ evaluates to 1; at this point EOL has been reached for this particle. The complexity of the algorithm is variable, as each particle must be simulated forward until its individual EOL, which is different for each particle.

Prediction requires hypothesizing future inputs of the system, $\hat{\mathbf{u}}_k$, because damage progression is rarely independent of the system inputs. In general, these inputs are unknown and must be chosen in order to satisfy both operational constraints and the type of prediction that is required. For the valve, we assume input-independence for the external leaks, as described by their damage models, but not for the remaining faults. For the valve, the problem of hypothesizing inputs is simplified. Each valve cycle corresponds to the same set of inputs, except for the fluid pressures p_L and p_R. However, these pressures can safely be assumed to be constant in our application domain, and, further, they have an almost negligible effect on valve behavior because the forces they produce are very small compared to the other forces acting on the valve. So, the future inputs for each valve cycle are deterministic. We can simply provide repeated valve cycles as input, and the prediction step will determine after how many valve cycles EOL will be reached for each particle.

Fig. 6 shows results from the prediction of friction damage for $N = 100$. Initially, the particles have a very tight distribution of friction parameter values, but the distribution of the wear parameter, w_r, is relatively large. As a result, the individual

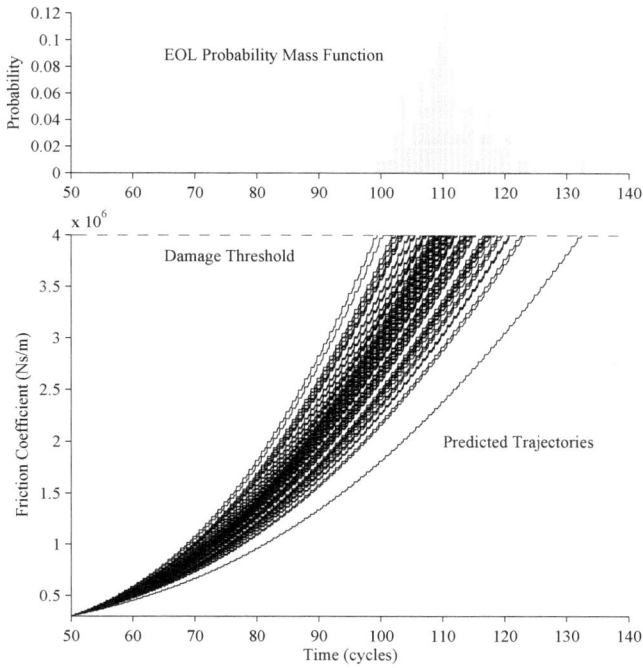

Figure 6. Prediction of friction damage

Parameter	Value
g	9.8 m/s^2
p_{atm}	1.01×10^5 Pa
p_{supply}	5.27×10^6 Pa
m	50 kg
r	6.00×10^3 Ns/m
k	4.80×10^4 N/m
x_o	2.54×10^{-1} m
L_s	3.81×10^{-2} m
A_p	8.10×10^{-3} m^2
V_{t_0}	8.11×10^{-4} m^3
V_{b_0}	8.11×10^{-4} m^3
ρ	7.10×10^2 kg/m^3
A_v	5.07×10^{-2} m^2
C_v	4.36×10^{-1}
R_g	2.96×10^2 J/K/kg
T	293 K
γ	1.4
Z	1
A_s	1.00×10^{-5} m^2
C_s	0.62

Table 1. Nominal valve parameters

trajectories for the different wear parameter values are easily distinguishable as EOL is approached. The different EOL values along with particle weights form an EOL distribution approximated by the probability mass function shown in the figure.

6 RESULTS

We apply the prognostics framework to a pneumatic valve of the Space Shuttle cryogenic refueling system. This system transfers cryogenic propellant (liquid hydrogen) from a storage tank to the vehicle tank through a network of pipes and valves. We focus on one of the transfer line valves in this system. In this section, we first develop and validate the model for the selected valve. Then, since only the discrete *open* and *closed* sensors are available for the valve, we establish, through a set of simulation-based experiments, that prognostics can still be performed using only those sensors. We then apply the prognostics approach to historical valve degradation data. Since proactive maintenance was performed early, no complete run-to-failure data exist. We augmented the missing portion with simulated data.

6.1 Model Validation

For the selected pneumatic valve, we identified model parameters using component specifications and estimated the remaining parameters (m, r, k, x_o, and C_s) from nominal data. The model parameter values are shown in Table 1. Note that the pneumatic gas is nitrogen.

Nominally, this valve opens in a little over 1 s and closes within 0.2 s. The valve has an opening time requirement of 5 s and a closing time requirement of 4 s, and these define functional thresholds for EOL. The valve is considered to have failed when these timing limits are exceeded. Further, the valve must fully close when actuating pressure is lost.

Due to the high safety margin of the fueling system, components must always meet tight operational constraints, and components are never intentionally run to failure. We focus on a particular time frame of historical data, consisting of 7 valve cycles, where a clear trend in performance degradation was observed before a maintenance action took place. We analyze this particular data set to determine the dominant damage mode and validate the corresponding damage model.

The obtained valve data includes only the *open* and *closed* sensors. Along with valve open/close commands, we may extract the following timing information, shown in Fig. 7:

1. $t_{open,1}$, defined as the difference between the time when the valve is commanded to open and it starts to move open,

2. $t_{open,2}$, defined as the difference between the time when the valve starts to move open and it fully opens,

3. $t_{open} = t_{open,1} + t_{open,2}$, which is the total time for the valve to open,

4. $t_{close,1}$, defined as the difference between the time when the valve is commanded to close and it starts to move closed,

5. $t_{close,2}$, defined as the difference between the time when the valve starts to move closed and it fully closes,

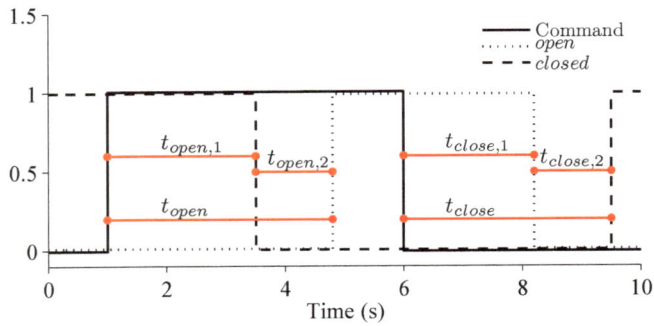

Figure 7. Pneumatic valve timing diagram. A command of 1 opens the valve, and a command of 0 closes the valve.

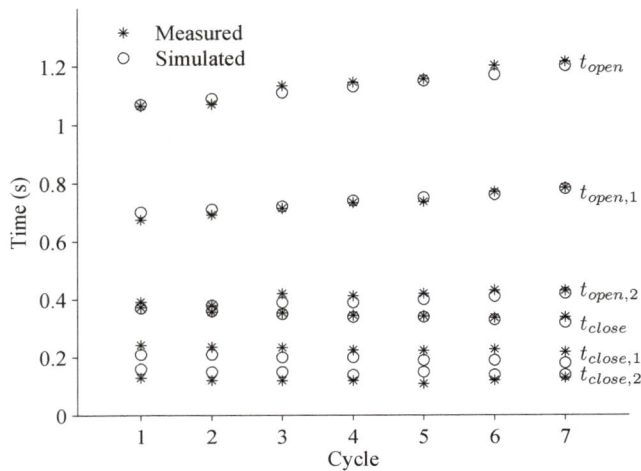

Figure 8. Pneumatic valve model comparison for growth of bottom external leak

6. $t_{close} = t_{close,1} + t_{close,2}$, which is the total time for the valve to close.

We plot these values for the degrading valve in Fig. 8. We note trends in each of the timing variables. Most importantly, the opening time increases, while the closing time decreases. From Fig. 4 (Section 3), we know that only two fault modes may cause this behavior, namely, the internal leak, and the bottom external leak. We may distinguish between these two damage modes by looking at $t_{open,1}$. For the internal leak, a slight decrease in this value is expected as the damage progresses, but for the bottom external leak, a significant increase in this value is expected. According to the data, this value increases, therefore, the bottom external leak is the only consistent damage mode, and must be dominant. Fig. 8 shows simulated timing values plotted alongside the measured timing values, with a wear rate of $w_{e,b} = 3.0 \times 10^{-7}$ m^2/cycle, which best matched the data. The simulated data captures all the trends of the real data, and the simulated timing values are all fairly close to the real values at each cycle.

6.2 Measurement Analysis

Since only the discrete *open* and *closed* sensors are available, we must first establish that prognostics can still be performed satisfactorily in this case. To investigate this issue, we performed a number of simulation experiments for each fault under different measurement sets. In each case, we used $N = 500$ particles, and tuned the random walk variances assuming that the order of magnitude of the wear parameters was known. Wear parameter values were selected so that EOL occurred near 100 cycles. For each fault and measurement set, 5 experiments were performed. Table 2 presents averaged results over these experiments. The performance metrics are defined in the Appendix. Estimation is evaluated based on the estimate of the unknown wear parameter, as it is this estimate that most greatly influences subsequent prognostics performance, since the future valve usage is well-defined. Estimation accuracy is calculated using the percentage root mean square error (PRMSE), where we ignore the first 10 cycles, which are associated with convergence. We calculate the spread using the relative median absolute deviation (RMAD), which is a robust measure of statistical dispersion. We compute also convergence of the wear parameter estimate, denoted as C_w, over the first 3 cycles. For a prediction point k_P, we compute measures of accuracy and precision. For accuracy, we use the relative accuracy (RA) metric (Saxena, Celaya, Saha, Saha, & Goebel, 2010). We use $\overline{RA}$ to denote the RA averaged over all prediction points. We calculate prediction spread using RMAD, which we denote as RMAD$_{RUL}$. We use $\overline{RMAD}_{RUL}$ to denote RMAD$_{RUL}$ averaged over all prediction points.

From Table 2, it is clear that prognostics can be performed successfully with all measurement sets, including using only the discrete sensors *open* and *closed*. Because of the loss of information in going from the continuous sensors to the discrete sensors, the state distribution has a much wider variance, usually increasing by around 50%. In turn, the wider variance causes the estimate to smooth out, and, as a result, PRMSE is lowered, and this corresponds to increases in RA, usually around 0.5%, however this is at the cost of a less confident prediction, as shown by RMAD$_{RUL}$. Further, convergence of the wear parameter estimate is slower using only the discrete sensors.

The results show that, if it is feasible to add the continuous sensors, prognostics performance can be improved significantly. The increase in information provided by these sensors leads to more confident predictions. Different sensors are more useful for different faults, so additional sensors may be selected based on which faults are more likely to appear, or more critical. For friction and spring damage, the position sensor is more useful than the pressure sensors, as shown by the decreased spread when $M = \{x\}$ compared to when $M = \{p_t, p_b\}$ for these faults. In contrast, the pressure sensors are more valuable for the leak faults.

Fault	M	PRMSE_w	$\overline{\mathrm{RMAD}}_w$	C_w	$\overline{\mathrm{RA}}_{RUL}$	$\overline{\mathrm{RMAD}}_{RUL}$
Friction	$\{x, f, p_t, p_b\}$	4.00	8.07	30.72	96.59	8.64
	$\{x\}$	3.92	9.38	28.79	97.24	8.62
	$\{p_t, p_b\}$	3.58	10.11	34.90	96.27	10.17
	$\{open, closed\}$	3.52	13.53	44.68	97.38	11.68
Spring Rate	$\{x, f, p_t, p_b\}$	5.58	10.87	32.70	94.91	11.04
	$\{x\}$	5.84	12.36	36.07	94.80	12.43
	$\{p_t, p_b\}$	5.47	13.14	31.74	94.51	13.09
	$\{open, closed\}$	5.10	16.58	43.50	94.58	16.40
Internal Leak	$\{x, f, p_t, p_b\}$	6.00	10.33	25.88	95.26	10.82
	$\{x\}$	5.84	13.81	21.40	95.29	14.37
	$\{p_t, p_b\}$	5.39	10.52	27.59	94.74	10.69
	$\{open, closed\}$	4.51	17.42	33.49	95.94	17.06
Top External Leak	$\{x, f, p_t, p_b\}$	5.33	9.27	21.54	95.73	9.32
	$\{x\}$	6.78	10.99	17.30	94.72	11.06
	$\{p_t, p_b\}$	4.00	10.04	20.83	96.87	10.12
	$\{open, closed\}$	5.14	14.11	18.82	95.94	14.24
Bottom External Leak	$\{x, f, p_t, p_b\}$	5.52	11.21	25.29	95.57	11.13
	$\{x\}$	6.02	14.49	25.15	95.04	14.22
	$\{p_t, p_b\}$	5.13	11.36	24.45	95.64	11.27
	$\{open, closed\}$	5.07	18.59	34.12	95.69	18.14

Table 2. Prognostics performance for different measurement sets

The valve considered here only sees discrete operation, where the valve is either fully closed or fully opened. In this case, the discrete open and closed sensors provide enough information for prognostics. For situations where the valve's position is controlled in a continuous manner, a continuous position sensor is necessary for prognostics. For such valves, this sensor is usually already available for feedback to the position controller.

6.3 Demonstration of the Approach

Given that prognostics may still be performed using only the *open* and *closed* sensors, we now apply the prognostic algorithms to the historical data. Because the actual data consisted of only 7 valve cycles before maintenance was performed, we extend the data with simulated valve data, based on the identified bottom external leak damage mode with $w_{e.b} = 3.0 \times 10^{-7}$ m^2/cycle, up to EOL at 38 cycles, at which the valve fails to open within the 5 second limit. Fig. 9 shows the valve timing data used for the demonstration. It is important to note that, even though the trend in valve opening time appears linear over the first 10 to 15 cycles, the overall trend is clearly nonlinear. This behavior is captured by the model, and serves as a justification of a model-based approach as opposed to simple trending strategies. Projecting a linear trend out from the 10th cycle would produce a severe overestimate of the true RUL.

To demonstrate a general solution, we allowed the particle

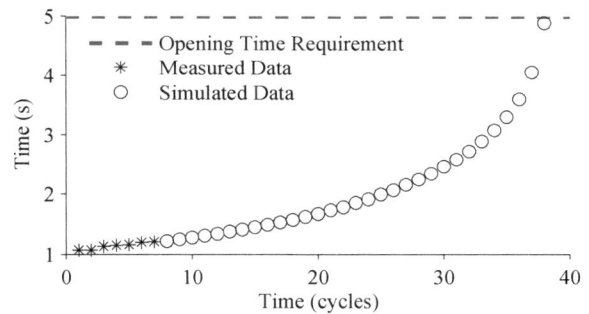

(a) Opening Times

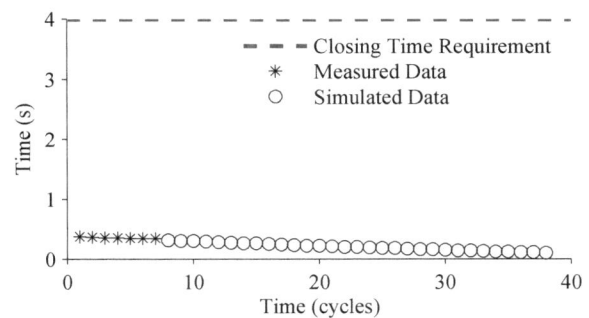

(b) Closing Times

Figure 9. Real and simulated valve timing

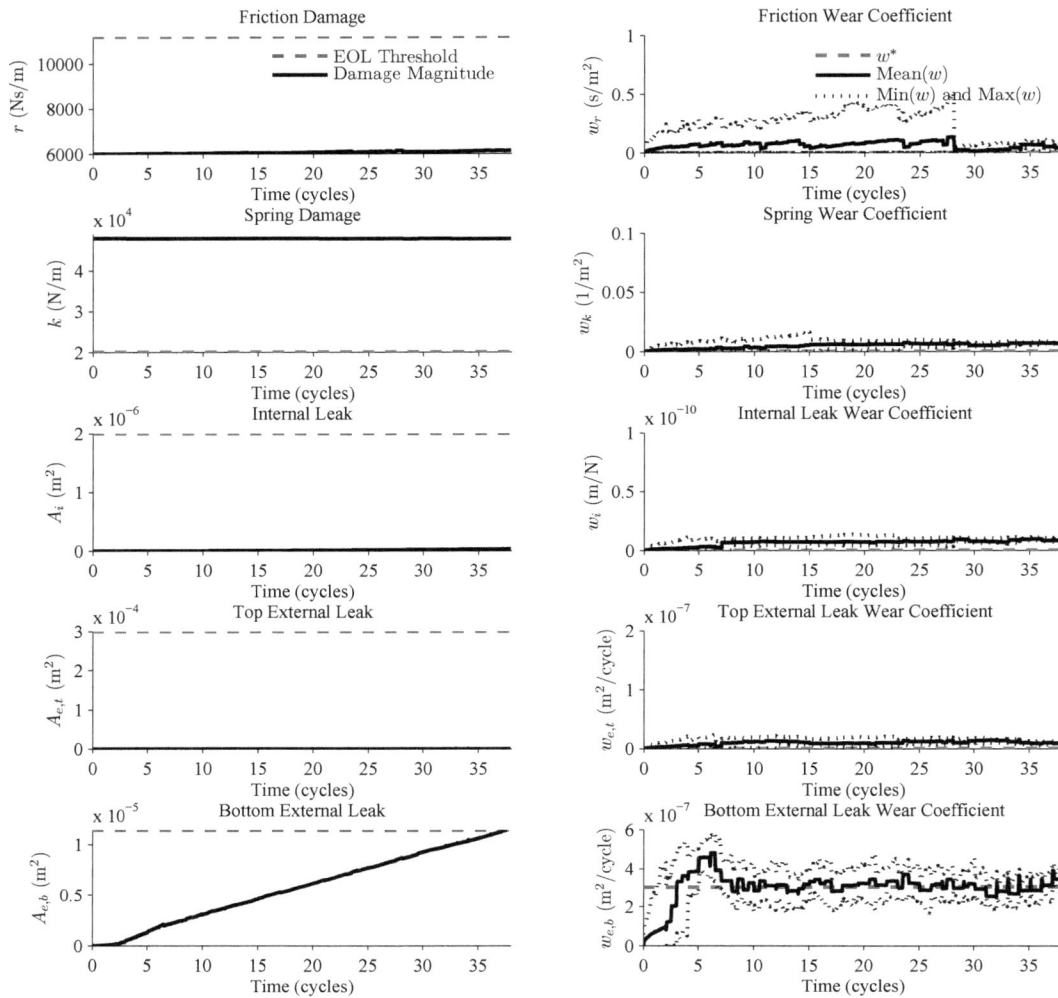

Figure 10. Estimation of pneumatic valve damage modes

filter to jointly estimate all the damage modes of the valve. Because of the large state space, $N = 2000$ was used. The algorithm should converge to correct wear rates for each damage mode, and, in this case, identify the bottom external leak as the dominant damage mechanism. Estimation results are shown in Fig. 10. The algorithm correctly identifies the bottom external leak as the dominant damage mode, since it is the only damage mode for which the estimated damage variable is significantly increasing. During the real data portion, the algorithm is still converging, and by the time the real data portion has ended, the algorithm estimates the wear parameter within 30% of the true value, and the mean RUL prediction is within 25% of the true RUL. The estimated value of the RUL at that time point is less than the true value, resulting in a conservative estimate. By 10 cycles, the mean RUL prediction has improved to within 7% of the true RUL.

The prognostics performance is summarized in Table 3. All times are given in cycles. RUL^* denotes the true RUL value.

We show the predictions starting at 5 cycles. At this point, the particle filter is still converging, but the RUL is predicted with an accuracy of 80%. Afterwards, RA ranges from 87 to 97%. The RMAD of the prediction varies from around 5 to 7%. So, RUL is predicted with reasonable accuracy, and the predictions are fairly confident. It is also useful to inspect the RUL prediction at a selected level of confidence. Fig. 11 shows the predictions at the 99% confidence level, i.e., the value at which 99% of the prediction distribution is greater than or equal to that value. The predictions at this confidence level always lie below the true RUL, therefore, they serve as conservative estimates upon which risk-averse decisions can be made.

Although encouraging, the results presented here represent only a limited validation of the overall approach, since the historical data covers only a single fault mode of the five considered, and extends only to 7 valve cycles. A more complete validation requires a more flexible testbed, such as one that

t_P	RUL^*	$\overline{RUL}$	RA_{RUL}	$RMAD_{RUL}$
5.0	33.00	26.60	80.60	7.28
7.5	30.50	27.53	90.27	5.49
10.0	28.00	26.04	92.99	7.65
12.5	25.50	24.42	95.77	5.96
15.0	23.00	21.75	94.57	7.61
17.5	20.50	21.84	93.46	6.04
20.0	18.00	16.55	91.93	6.52
22.5	15.50	13.53	87.29	6.63
25.0	13.00	11.64	89.52	7.87
27.5	10.50	10.15	96.71	6.90
30.0	8.00	6.78	84.74	5.14
32.5	5.50	5.34	97.11	4.81
35.0	3.00	2.93	97.69	7.16

Table 3. Pneumatic valve prognostics performance

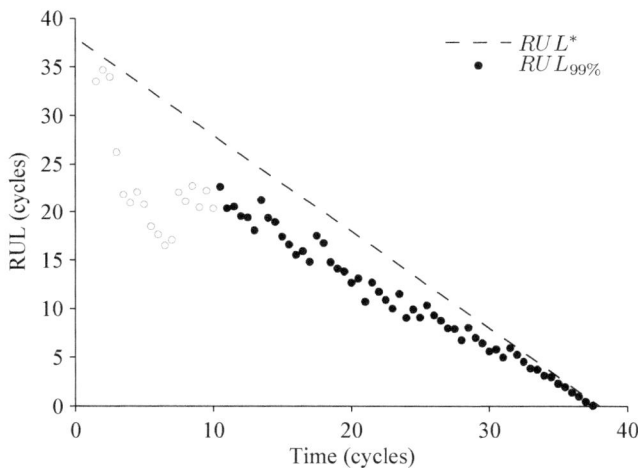

Figure 11. RUL predictions at the 99% confidence level

would allow the injection of faults and/or valves to be run to failure.

7 RELATED WORK

Very little work has been done in valve prognostics. A health monitoring application for a pneumatic valve in a pressure control system is considered in (Gomes, Ferreira, Cabral, Glavão, & Yoneyama, 2010). Friction damage, spring aging, and internal leaks are also identified as common failure modes in their application. The distribution of pressure signals is monitored using a probability integral transform, and a measure of dissimilarity from a baseline distribution serves as a health index. The approach is purely data-driven, relying on a nominal distribution of pressure signals to serve as a baseline, and comparing future pressure distributions to the baseline. No identification of the form of damage present is performed, and prediction is not performed either.

Model-based prognostic techniques have been applied to other systems, and particle filtering techniques have been particularly popular. In (Saha & Goebel, 2009), a particle filtering approach is applied to end of discharge and end of life prediction for lithium-ion batteries. In (Orchard et al., 2008), a particle filter-based diagnosis and prognosis framework is applied to crack growth in aircraft components. They also implement outer correction loops to control the variance of the random walk of the unknown parameters based on prediction error. This type of approach is preferred to the fixed variance method employed in this paper, however, the approach presented in (Orchard et al., 2008) is only applicable when a feature can be derived that is a direct function of a single fault dimension, which is not the case with valves. In (Abbas, Ferri, Orchard, & Vachtsevanos, 2007), the authors apply a particle filter-based prognosis method to prediction of battery grid corrosion. A physics-based methodology to prediction of aircraft engine bearing spalls is developed in (Bolander, Qiu, Eklund, Hindle, & Rosenfeld, 2010). Particle filters are used for estimation of spall size, using diagnostic information from vibration and oil debris sensors in the update step. Prediction is performed using anticipated future load and speed, and the approach was evaluated in a physical testbed. Although the size of our state-parameter vector is much larger than those considered in these previous works, our particle filter-based approach does not differ from these previous approaches in any fundamental way.

Other filtering techniques may also be used. In (Chelidze, 2002), the system is modeled in a hierarchical fashion, with a fast-time observable subsystem coupled to a slow-time damage subsystem, where damage is tracked using a Kalman filter. The approach is applied to tracking open circuit battery voltages in a vibrating beam testbed using only beam displacement measurements. In a related approach, a model-based prognosis methodology is developed in (Luo et al., 2008) using an interacting multiple model filter for state-parameter estimation and prediction. The approach is applied to prognosis of fatigue cracks in a simulation of an automotive suspension system.

8 CONCLUSIONS

In this paper, we developed a model-based prognostics approach using particle filters for joint state-parameter estimation. The estimated damage state of the valve is propagated forward in time to predict EOL and form EOL and RUL distributions. We applied the approach to a pneumatic valve, developing a detailed physics-based model that included damage progression models. Through simulation experiments, we established that prognostics could be performed successfully using only the discrete position sensors of the valve, and applied the approach to real valve data under this constraint. The results here demonstrated the effectiveness of a model-based approach, and gave insight into the selection of sensors for valve prognostics.

Here, we advocate a model-based approach, where performance will depend greatly on the accuracy of the model provided. Clearly, developing such models is a key obstacle. The model developed here is suitable for a normally-closed pneumatic valve with a piston-based actuation mechanism. With small modifications, the model may also be used for normally-open valves or valves that use a diaphragm mechanism. With other valve types, the actuation mechanism changes, so a new model must be developed. However, there will be several commonalities between the different models. For instance, in past work, we have applied similar modeling techniques to a solenoid valve (Daigle & Goebel, 2010a). Here, damage modes for the spring and friction are also present, and the same damage progression models are used. The damage models can also be translated into a forms for rotational motion, as in (Daigle & Goebel, 2011), where a modified friction damage progression is used in a centrifugal pump model. Of course, one must also consider the cost of developing such detailed models against the corresponding prognostics performance achieved, and this is the subject of recent work (Daigle et al., 2011).

The valves considered here are always operated in the same way; they fully open, then fully close. Therefore, future valve inputs can be easily hypothesized, and in this paper we assumed a single known future input trajectory to provide EOL and RUL in the number of valve cycles. In general, the future inputs may not be well-known, and different sets or distributions of possible inputs must be considered. In future work, we will extend the approach to these cases. Further, although the algorithm assumes all damage modes may be progressing simultaneously, the results here cover cases where only a single damage mode is progressing. Recent work considers the case when multiple damage modes are active for a centrifugal pump simulation (Daigle & Goebel, 2011), and investigating this for pneumatic valves, especially with limited sensing, where fault masking becomes an issue, is considered as future work. Applying the approach to additional testbeds is also part of future work.

APPENDIX

Performance Metrics

Estimation performance is evaluated based on the estimate of the unknown wear parameter. Estimation accuracy is calculated using the percentage root mean square error (PRMSE), which expresses relative estimation accuracy as a percentage:

$$\text{PRMSE}_w = 100 \sqrt{\text{Mean}_k \left[\left(\frac{\hat{w}_k - w_k^*}{w_k^*} \right)^2 \right]},$$

where $\hat{w}_k$ denotes the estimated wear parameter value at time k, w_k^* denotes the true wear parameter value at k, and Mean_k denotes the mean over all values of k. In computing PRMSE,

we typically ignore the initial portion of data associated with convergence.

We calculate the spread using the relative median absolute deviation (RMAD), which expresses the spread relative to the median as a percentage:

$$\text{RMAD}(X) = 100 \frac{\text{Median}_i \left(|X_i - \text{Median}_j(X_j)| \right)}{\text{Median}_j(X_j)},$$

where X is a data set and X_i is an element of that set. For estimation spread, for time k we use the distribution of wear parameter values given by the particle set at k as the data set. We summarize RMAD over an experiment by averaging RMAD over all k. We denote the average RMAD over multiple k using $\overline{\text{RMAD}}_w$:

$$\overline{\text{RMAD}}_w = \text{Mean}_k(\text{RMAD}_{w,k}),$$

where $\text{RMAD}_{w,k}$ denotes the RMAD of the wear parameter at time k.

The final estimation metric is convergence of the wear parameter estimate, denoted as C_w. We use the definition of the convergence metric described in (Saxena et al., 2010), where the convergence of a curve is expressed as the distance from the origin to the centroid under the curve (where a shorter distance is better). We use the absolute error of the hidden parameter estimate as the curve. We compute convergence only over the initial convergence period, so that errors after convergence do not contribute. The same time window must be used consistently over all experiments to properly compare.

For a prediction point k_P, we compute measures of accuracy and precision. For accuracy, we use the relative accuracy (RA) metric (Saxena et al., 2010), computed as a percentage:

$$\text{RA}_{k_P} = 100 \left(1 - \frac{|RUL_{k_P}^* - \text{Median}_i(RUL_{k_P}^i)|}{RUL_{k_P}^*} \right),$$

where $RUL_{k_P}^*$ is the true RUL at time k_P. Here, we have chosen the median as a robust measure of central tendency of the distribution. We use $\overline{\text{RA}}$ to denote the averaged RA over all prediction points. We calculate prediction spread using RMAD, which we denote as RMAD_{RUL}. We average RMAD_{RUL} over all prediction points to obtain a single number representing prediction spread over a single experiment, denoted using $\overline{\text{RMAD}}_{RUL}$.

NOMENCLATURE

EOL	end of life
RUL	remaining useful life
t	time (continuous)
k	time (discrete)
t_P or k_P	time of prediction
$\mathbf{x}$	state vector
$\boldsymbol{\theta}$	parameter vector
$\mathbf{v}$	process noise vector
$\mathbf{y}$	output vector
$\mathbf{n}$	measurement noise vector
T_{EOL}	EOL threshold
PRMSE	percent root mean square error
RMAD	relative median absolute deviation
RA	relative accuracy

REFERENCES

Abbas, M., Ferri, A. A., Orchard, M. E., & Vachtsevanos, G. J. (2007). An intelligent diagnostic/prognostic framework for automotive electrical systems. In *2007 IEEE Intelligent Vehicles Symposium* (pp. 352–357).

Arulampalam, M. S., Maskell, S., Gordon, N., & Clapp, T. (2002). A tutorial on particle filters for on-line nonlinear/non-Gaussian Bayesian Tracking. *IEEE Transactions on Signal Processing, 50*(2), 174–188.

Bolander, N., Qiu, H., Eklund, N., Hindle, E., & Rosenfeld, T. (2010, October). Physics-based remaining useful life prediction for aircraft engine bearing prognosis. In *Proceedings of the Annual Conference of the Prognostics and Health Management Society 2010.*

Byington, C. S., Watson, M., Edwards, D., & Stoelting, P. (2004, March). A model-based approach to prognostics and health management for flight control actuators. In *Proceedings of the 2004 IEEE Aerospace Conference* (Vol. 6, pp. 3551–3562).

Cappe, O., Godsill, S. J., & Moulines, E. (2007). An overview of existing methods and recent advances in sequential Monte Carlo. *Proceedings of the IEEE, 95*(5), 899.

Chelidze, D. (2002). Multimode damage tracking and failure prognosis in electromechanical system. In *Proceedings of the SPIE Conference* (Vol. 4733, pp. 1–12).

Clapp, T. C., & Godsill, S. J. (1999). Fixed-lag smoothing using sequential importance sampling. *Bayesian Statistics IV*, 743–752.

Daigle, M., & Goebel, K. (2009, September). Model-based prognostics with fixed-lag particle filters. In *Proceedings of the Annual Conference of the Prognostics and Health Management Society 2009.*

Daigle, M., & Goebel, K. (2010a, October). Improving Computational Efficiency of Prediction in Model-based Prognostics Using the Unscented Transform. In *Proceedings of the Annual Conference of the Prognostics and Health Management Society 2010.*

Daigle, M., & Goebel, K. (2010b, March). Model-based prognostics under limited sensing. In *Proceedings of the 2010 IEEE Aerospace Conference.*

Daigle, M., & Goebel, K. (2011, March). Multiple damage progression paths in model-based prognostics. In *Proceedings of the 2011 IEEE Aerospace Conference.*

Daigle, M., Roychoudhury, I., Narasimhan, S., Saha, S., Saha, B., & Goebel, K. (2011, September). Investigating the Effect of Damage Progression Model Choice on Prognostics Performance. In *Proceedings of the Annual Conference of the Prognostics and Health Management Society 2011.*

Doucet, A., Godsill, S., & Andrieu, C. (2000). On sequential Monte Carlo sampling methods for Bayesian filtering. *Statistics and Computing, 10*, 197–208.

Gomes, J. P. P., Ferreira, B. C., Cabral, D., Glavão, R. K. H., & Yoneyama, T. (2010, October). Health monitoring of a pneumatic valve using a PIT based technique. In *Proceedings of the Annual Conference of the Prognostics and Health Management Society 2010.*

Hutchings, I. M. (1992). *Tribology: friction and wear of engineering materials.* CRC Press.

Julier, S., & Uhlmann, J. (1997). A new extension of the Kalman filter to nonlinear systems. In *Proceedings of the 11th International Symposium on Aerospace/Defense Sensing, Simulation and Controls* (pp. 182–193).

Kitagawa, G. (1996). Monte Carlo filter and smoother for non-Gaussian nonlinear state space models. *Journal of Computational and Graphical Statistics, 5*(1), 1–25.

Liu, J., & West, M. (2001). Combined parameter and state estimation in simulation-based filtering. *Sequential Monte Carlo methods in Practice*, 197–223.

Luo, J., Pattipati, K. R., Qiao, L., & Chigusa, S. (2008, September). Model-Based Prognostic Techniques Applied to a Suspension System. *IEEE Transactions on Systems, Man and Cybernetics, Part A: Systems and Humans, 38*(5), 1156-1168.

Orchard, M. (2007). *A Particle Filtering-based Framework for On-line Fault Diagnosis and Failure Prognosis.* Unpublished doctoral dissertation, Georgia Institute of Technology.

Orchard, M., Kacprzynski, G., Goebel, K., Saha, B., & Vachtsevanos, G. (2008, October). Advances in uncertainty representation and management for particle filtering applied to prognostics. In *Proceedings of International Conference on Prognostics and Health Management.*

Perry, R., & Green, D. (2007). *Perry's chemical engineers' handbook.* McGraw-Hill Professional.

Roemer, M., Byington, C., Kacprzynski, G., & Vachtsevanos, G. (2005). An overview of selected prognostic technologies with reference to an integrated PHM architecture. In *Proceedings of the First International Forum on Integrated System Health Engineering and Manage-*

ment in Aerospace.

Saha, B., & Goebel, K. (2009, September). Modeling Li-ion battery capacity depletion in a particle filtering framework. In *Proceedings of the Annual Conference of the Prognostics and Health Management Society 2009.*

Saha, B., & Goebel, K. (2011). Model Adaptation for Prognostics in a Particle Filtering Framework. *International Journal of Prognostics and Health Management.*

Saxena, A., Celaya, J., Saha, B., Saha, S., & Goebel, K. (2010). Metrics for Offline Evaluation of Prognostic Performance. *International Journal of Prognostics and Health Management (IJPHM)*, *1.*

Schwabacher, M. (2005). A survey of data-driven prognostics. In *Proceedings of the AIAA Infotech@Aerospace Conference.*

INTERNATIONAL JOURNAL OF PROGNOSTICS AND HEALTH MANAGEMENT, VOL.2 (2011)

Model Based Approach for Identification of Gears and Bearings Failure Modes

[1]Dr. Renata Klein, [2]Eduard Rudyk, [3]Dr. Eyal Masad, [4]Moshe Issacharoff

[1-4]*R.K. Diagnostics, Misgav Industrial Park, P.O.B. 66, D.N. Misgav 20179, Israel*

Renata.Klein@RKDiagnostics.co.il
Eddie.Rudyk@RKDiagnostics.co.il
Eyal.Masad@RKDiagnostics.co.il
Moshe.Issacharoff@RKDiagnostics.co.il

ABSTRACT

This paper describes the algorithms that were used for analysis of the PHM'09 gear-box. The purpose of the analysis was to detect and identify faults in various components of the gear-box. Each of the 560 vibration recordings presented a different set of faults, including distributed and localized gear faults, typical bearing faults and shaft faults. Each fault had to be pinpointed precisely.

In the following sections we describe the algorithms used for finding faults in bearings, gears and shafts, and the conclusions that were reached. A special blend of pattern recognition and signal processing methods was applied.

Bearings were analyzed using the orders representation of the envelope of a band pass filtered signal and an envelope of the de-phased signal. A special search algorithm was applied for bearing feature extraction. The diagnostics of the bearing failure modes was carried out automatically. Gears were analyzed using the order domains, the quefrency of orders, and the derivatives of the phase average.

1 PHM' 09 CHALLENGE

The PHM09 data set included 560 recordings of 2 seconds, measured on the gearbox described in Figure 1, using two vibration sensors and a tachometer. All the bearings were similar. Some of the signals were recorded when the gearbox was in 'spur' configuration, and others when it was in 'helical' configuration. Data were collected at 30, 35, 40, 45 and 50 Hz shaft speed, under high and low loading.

The PHM'09 data presented a few specific difficulties:

To begin with, our automatic standard approach is based on the existence of a robust "baseline" – that is a set of characteristic signature statistics derived from undamaged

units in typical operating modes. The algorithms compare between the baseline and the signature of the unit under test. The fact that in the challenge data, the "undamaged" cases were unidentified complicated the generation of a trusted baseline and the optimization of both false alarms rate and misdetections rate.

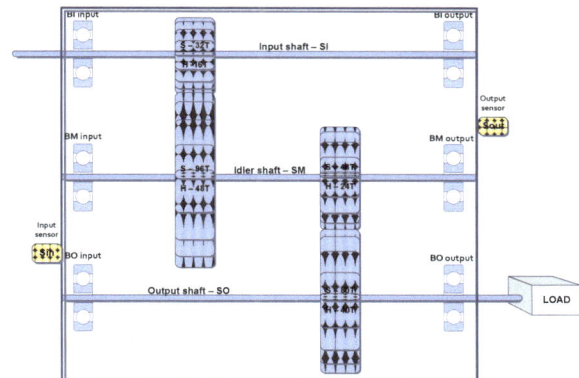

Figure 1: Challenge apparatus: spur (S) and helical (H) configurations

The other difficulty resulted from the short duration of the recordings. It limited the performance of the signal processing algorithms in enhancing failure modes manifestations and in separating between excitations of different components. In the challenge apparatus, the separation between the different components manifestations was especially challenging due to the adjacency of the bearing shock rates to the shaft harmonics and gear sidebands, and due to the gear ratios that have generated overlapping pointers (pointer – a frequency characteristic to a specific component).

2 BEARING ANALYSIS

The signatures that may reveal the bearing failure modes manifestation are the power spectral density – PSD (see Figure 13) and the orders representation of the envelope. Both are in the orders domain, i.e. the frequency equivalent domain after synchronization of the vibrations with the rotating speed (resampling according to one of the shafts). All the signatures used for bearing diagnostics were in the orders domain according to the input shaft.

The most significant signatures used for bearing analysis were orders representations of two types of envelopes, namely envelope of a band pass filtered signal and the envelope of the de-phased signal (R. Klein et. al.). As can be observed in Figure 13 in the orders of the envelope of the de-phased signal the peaks related to the bearing failure modes are visible while in the regular orders representation the peaks are masked by the other rotating components excitation of higher energies.

2.1 Envelope of the filtered signal

An elevated frequency region at 7-10.5 KHz (in most cases separated from the high harmonics of the gear peaks) was identified as enhancing the bearing tones.

The identification of the frequency region was based on the visual inspection of the frequency PSD and the kurtogram of the wavelets decomposition of the raw signals. More details can be found in Sawalhi et. al., Antoni et. al., and Yang et. al. (2004).

The signal was band pass filtered around these frequencies, and then resampled according to the input shaft rotating speed. The envelope $e(t)$ of the signal $x(t)$ is the absolute value of the analytic signal $e(t) = |x(t) + jH\{x(t)\}|$ where H denotes the Hilbert transform. The spectrum of the squared envelope was estimated in order to reveal the repetition rate of the shocks.

2.2 Envelope of the de-phased signal

The de-phase algorithm removes the phase averages (synchronous average) of all the shafts from the resampled signal, isolating the asynchronous excitations in the vibrations signal. The demodulated asynchronous excitations are represented by the spectra of the envelope of the de-phased signal (Klein et. al.).

The de-phase process was performed in 5 stages (see Figure 2):

1. Generation of the phase average of the idler shaft (SM);

2. Generation of the phase average of the output shaft (SO);

3. Filtration of harmonics of the input shaft out from the phase average of the output shaft using an ideal FFT band stop comb filter (removal of the harmonics of the input shaft coinciding with the multiples of 5 of the output shaft);

4. Resampling of the raw data according to the input shaft (SI);

5. Removal of the phase averages from the resampled data.

The envelope was generated using the Hilbert transform and its spectrum was calculated.

The quality of the signatures obtained depends on the capability to isolate and remove the synchronous vibrations excitations, i.e. the quality depends on the phase average quality. In the PHM case we had some factors which made it easier to estimate the phase average signals, namely an accurate RPS measurement and vibrations acquisition during steady states. The main problems were due to the fact that the record length that were too short (especially considering the low RPS recordings) to allow enough averages for excitation sources separation.

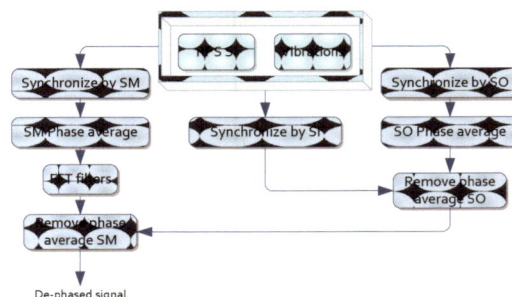

Figure 2. The de-phase process

2.3 Bearing tones

Each time a contact with the damaged bearing surface occurs a shock is excited.

The repetition rate of the shocks with respect to the shaft rotating speed represents the rate of contacts of the damaged surface with other rolling elements. The rate of contacts on each type of surface (outer race, inner race, rolling elements, or cage) depends on the bearing geometry and the rotating speed of the inner and/or outer race. These rates of shocks characterize the damaged bearing and its failure mode. Assuming no slip between surfaces, the kinematic rates of contacts, i.e. Ball Pass Frequency Inner Race (BPFI), Ball Pass Frequency Outer Race (BPFO), Ball Spin Frequency (BSF), and the Fundamental Train Frequency (FTF), were calculated (see Table 1, Klein et. al., and McFadden et. al. (1990, 1984)).

The radial load on the bearing determines the strength of the impact created when an element rolls over a fault. Due to load variations during the rotation the repetitive shocks amplitudes are modulated. A fault in a stationary bearing race will be subjected to a constant force at each roll and consequently will not be modulated. A fault in a rotating race will be subjected to a varying force repeating itself with the RPS of the race. Thus, the pulse train will be amplitude-modulated with the RPS of the race. The frequency caused by a ball fault, will be amplitude-modulated by the RPS of the cage (FTF) and the rate of contacts with the damaged ball will be twice the ball spin frequency (BSF).

The pointers by SI orders (multiples of the rotating speed of the input shaft) for the bearings are listed in Table 1, including pairs of bearings on the input shaft (BI), bearings on the idler shaft (BM), and bearings on the output shaft (BO).

	BPFI	RPS	BPFO	BSF	2BSF	FTF
BI	4.948	1	3.052	1.992	3.984	0.382
BM	1.648	1/3	1.019	0.664	1.329	0.042
BO	0.989	0.2	0.611	0.399	1.797	0.015

Table 1: Bearing tones and pointers in orders of the input shaft (SI)

In order to differentiate between the bearing tones and the other peaks excited by other rotating components (gears and shafts) and in order to emphasize the expected patterns of peaks generated by damaged bearings the highest possible resolution is required.

2.4 High resolution spectra

As can be seen in Table 1, the separation of BO bearing tones would require a resolution of ~0.005 SI-orders, the separation of BM bearing tones would require a resolution of ~0.01 SI-orders, and the separation of BI bearing tones would require ~0.02 order by SI. With records of 2 sec. the maximum resolution that can be obtained while keeping a reasonable SNR (averaging enough FFT frames) is of 0.031 SI-orders.

The resolution of the spectra, Welch's PSD estimate – P_{xx} in Eq. (1), was improved using a procedure developed especially for the PHM challenge allowing the averaging of several (N) periodograms, $|X_i|$ in Eq. (1), calculated with a resolution of 0.016 SI-orders. Instead of dividing the time series data into overlapping segments, computing a modified periodogram of each segment, and then averaging the PSD estimates, the averaging was performed on periodograms corresponding to different records that were classified as similar.

$$P_{xx} = \frac{1}{N} \sum_{i=1}^{N} |X_i|^2 \qquad (1)$$

The similar records have been determined based on the results of hierarchical clustering. The algorithm for clustering of similar recordings had as inputs the orders by SI representation and the orders of both envelopes, of the filtered signal and of the de-phased signal. All the recordings having cophenetic distances below 65 were considered similar. The results of the hierarchical clustering were validated by other similarity checks and visual inspection.

With the high resolution obtained the capability to isolate the damaged bearings on input shaft (SI) and in some cases on the idler shaft (SM) was improved. The comparison of results for a specific recording with a damaged bearing on the input shaft is illustrated in Figure 12. In the high resolution spectrum the peaks representing the faulty bearing (BPFI and the corresponding sidebands) have higher levels and are separated from the integer orders enabling the distinction of sidebands of different harmonics of BPFI.

2.5 Bearing features

The bearing tones derived from the bearing geometry may be inaccurate due to variations in the contact angle (axial loading) and due to slippage. In the orders domain, when the bearing tones are shifted the entire pattern of each harmony is shifted while the sidebands remain at the same distances. Therefore a special algorithm that searches and identifies the pointer location by pattern was applied. The algorithm for bearing feature extraction is using a reference signature (the respective baseline) in order to determine the location of the bearing's pointers.

The algorithm applies a comb filter representing the searched pattern (applied for each failure mode separately) and calculates the most probable location of the bearing tone and the median of the Mahalanobis distance of the entire pattern from the baseline population. The energies of the peaks above the background are then calculated completing the feature extraction for bearings.

Initially a reference signature was created for each group of recordings at a certain nominal rotating speed based on the 15th percentile of the levels of the orders representations of the envelopes. Using this reference, the features of bearing failure modes have been extracted, and several recordings without damaged bearings were selected at each nominal rotating speed. In the second stage groups of 8-16 recordings were selected at each rotating speed and the baseline signature statistics were estimated (see Figure 3).

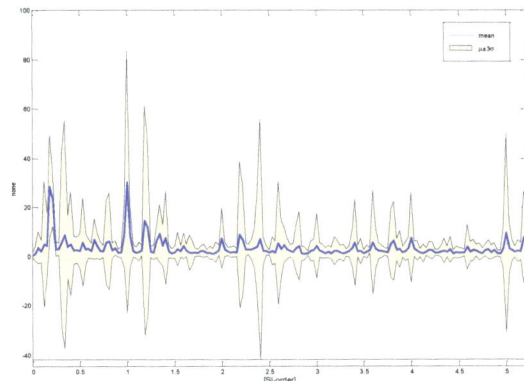

Figure 3. Baselines for bearings

The results of the feature extraction of bearings on the input shaft are displayed in Figure 4 and the results for bearings on the idler shaft are displayed in Figure 5. The figures represent the scores of the different failure modes of the bearings (the median of the Mahalanobis distance of the peaks with respect to the baseline). The scores of the bearings on the output shaft are not displayed because no damaged bearings have been detected.

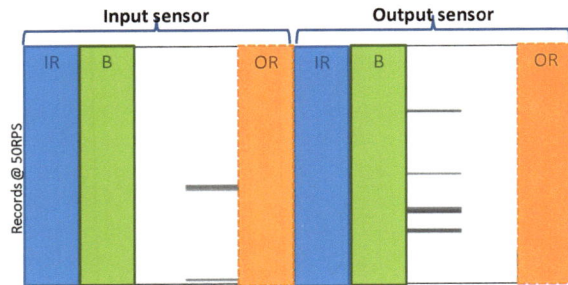

Figure 4. Scores of different failure modes of input shaft bearings, IR – inner race fault, OR – outer race fault, B – ball fault

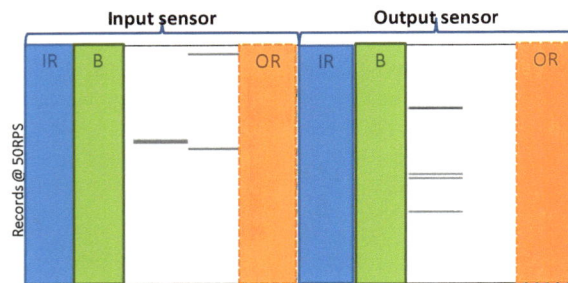

Figure 5. Scores of different failure modes of idler shaft bearings, IR – inner race damage, OR – outer race damage, B – ball damage

The algorithm detected only inner race and outer race damages of the bearings on the input shaft and only outer race damages of the bearings on the idler shaft. In several cases the same failure mode was detected on both sensors.

2.6   Bearing Decisions

The reasoning behind bearing decisions was similar for all bearings and failure modes.

In the cases where the scores (Mahalanobis distances from the baseline) were exceptionally high both in the envelope of the filtered signal and in the envelope of the de-phased signal and the pointer location was not an exact multiplier of one of the shaft speeds, we calculated the total energy of all the pointers in the pattern over 2-3 harmonies (including sidebands if relevant). The energies were extracted from the PSD of the envelope of the de-phased signal.

The total energies calculated per sensor were compared and in cases of clear discrimination the respective higher energy side (input side or output side) was diagnosed as having a

damaged inner/outer race or ball. The levels comparison is based on the assumption that the transmission path from each bearing to the adjacent sensor is similar. This assumption may not hold in practice.

In some recordings we observed very high levels of bearing tones on both sensors. These may be generated by combinations of faults, e.g. a bent or unbalanced input shaft and a damaged bearing, or two damaged bearings on the input shaft, or even high loads. We assumed that both bearings on the same shaft cannot have the same failure mode. In order to discriminate between these cases marked examples with faulty bearings are required for thresholds tuning.

In some cases peaks were observed and detected by the feature extraction algorithm at exact multiples of the shaft speeds (up to one point of the spectrum resolution). These cases were ignored in the decision process because they may be generated by modulation of faulty gears. The capability of both the filter and the de-phase algorithms to separate between the gears and bearings excitations was imperfect in this case: the de-phase algorithm capabilities were limited due to the short recordings (especially at low rotation speeds), and the envelope of the filtered signal was contaminated by faulty gears effects (at high rotation speeds). It seems, though, that the de-phase envelope has better capability to isolate the bearing excitations.

Another unknown was whether it was correct to ignore the patterns observed with low levels. When the sensor is located close to the damaged bearing and the excitations from other rotating components are relatively low, it can be that the bearing tones may be observed even when the bearing is in good condition. However, since we do not have such experience (maybe due to the complexity of the machinery that we usually diagnose), we did not know if that is the case here.

3   GEAR ANALYSIS

The most widely used signatures in gears analysis are in the orders, quefrency of orders and cycle domains (respectively equivalent to frequency, quefrency and time domains). The time history is mapped to the cycles' domain after synchronization (resampling) according to the rotating speed of a shaft.

The cepstrum of the orders representation was generated. The cepstrum reflects the repetition rate of sidebands (due to frequency modulation) and their average level in several peaks in the quefrency of orders domain (Zacksenhouse et. al., Antoni et.al.).

The separation of the vibrations of one gearwheel is achieved by calculating the phase average according to the respective shaft. Averaging is applied to enhance deterministic effects that are synchronized with the rotation of the relevant shaft. The phase average signal reveals the vibration induced by the meshing of each tooth on the relevant gear.

The phase average removes the asynchronous components by averaging the resampled signal in each cycle of rotation. All the signal elements that are not in phase with the rotation speed are eliminated, leaving the periodic elements represented in one cycle, i.e. the elements corresponding to harmonics of the shaft rotating speed.

Phase average with frequency f is designed to remove elements in V which are not periodic with the period $N=1/f$.

$$y_n := \frac{1}{M} \sum_{m=0}^{M-1} v_{n+mN} \quad n = 1, \ldots, N \quad (2)$$

Note that y is a vector in $\mathbf{R}^N$ representing a single cycle.

The phase average capability to filter out the asynchronous elements depends on the number of cycles averaged (in this case M). Therefore, it is preferable to average over as many cycles as possible, i.e. over a long period of time. In the case of the PHM recordings the number of cycles averaged differed pending on the shaft rotating speed. For the input shaft the number of averages was 60-100 depending on the rotating speed. The number of averages for the idler shaft was in the range 20-33 and for the output shaft it was in the range of 12-20 (for SI RPS of 30-50Hz respectively). Therefore, the capability to filter out the asynchronous elements (in our case the bearing effects) was limited and the phase averages were more noisy than usual. In Figure 6 we can see a sample of phase average signatures filtered around the first harmonic of the gearmesh of a wheel (idler shaft at the output side – spur gears configuration).

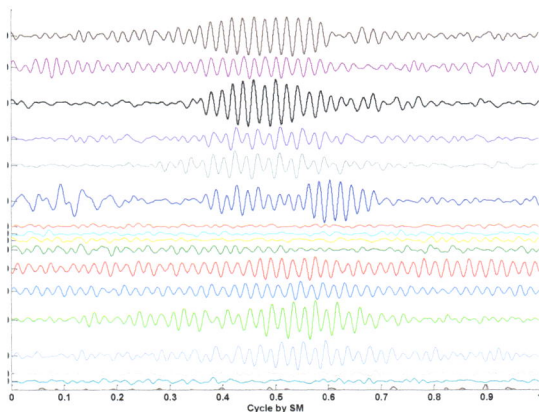

Figure 6. Examples of phase averages filtered around the first harmonic of the spur output gear (SM cycle)

Detection of abnormal meshing of individual teeth was achieved by further processing of the averaged signal into three types of signals: regular, residual, and difference (Zacksenhouse et. al.). The regular signals are obtained by passing the phase averaged signal through a multi-band-pass filter, with pass-bands centered at the meshing frequency and its harmonics (1÷5). It is essentially the cycle-domain average of the vibrations induced by a single tooth. The residual signals are obtained by removing the meshing frequency harmonics. The difference signals are obtained by removing the meshing frequencies and adjacent sidebands, i.e. separating between the frequency and amplitude modulations.

The envelope and the unwrapped phase of the regular, difference and residual parts of all the harmonics describe the characteristics of the amplitude modulation.

3.1 Gear Features

Generally, the phase average is dominated by the gear meshing components and some low-order amplitude modulation and/or phase modulation components. These modulation effects are generated by transmission errors related to geometric and assembly errors of the gear pairs. When a localized gear fault is present, a short period impulse will appear in each complete revolution. This produces additional amplitude and phase modulation effects. Due to its short period nature, the impulse produces high order low-amplitude sidebands surrounding the meshing harmonics in the spectrum. The removal of the regular gear meshing harmonics (residual part) sometimes with their low-order sidebands (difference part) from the phase average emphasizes the portion predominantly caused by gear fault and geometrical and assembly errors. Statistical measures of the residual part were used to quantify the fault-induced shocks.

Orders representation of the phase average according to each shaft was generated. The peak values at each harmonic of the shaft rotating speed representing gearmesh orders and sidebands were stored as features characterizing the gears.

The features of gears were extracted both for the spur and helical configurations. The features extracted for each gear wheel included: the even un-normalized statistical moments (RMS, kurtosis, 6th and 8th moments) of the regular, residual and difference parts, the even moments of the envelope of the regular, residual and difference, the peak values of the orders representation of the phase average for the gearmesh and the respective shaft sidebands, and the levels of the cepstrum representing the quefrency of the respective shafts.

In the regular procedure these features are "normalized" to the baseline transforming the moments into the known features FM0, FM2, FM4, FM6, FM8, NA4, NB4, etc. (see Lebold et. al.) and in later stages the gears wheels scores are determined using Mahalanobis distances from the baseline features. This procedure was modified for the case of the challenge when the normalization process and the distances measurement were omitted.

3.2 Spur/ Helical Clustering

It is well known that spur gears generate higher vibrations levels compared to helical gears. Therefore one of the features selected for the gearbox configuration classification was the energy of the background of the frequency spectra. The background signature was selected in order to avoid influence of the damaged components. The background of the frequency spectra was estimated based on an adaptive clutter separation algorithm.

The other features used for classification included the ratios of the energies of the spur/helical gearmesh orders of the wheel on the output shaft and of the wheels on the idler shaft. Recordings with similar rotating speed were clustered (hierarchical clustering) and classified using the above features as spur or helical.

3.3 Gear Diagnostics

The numerous features extracted for each gear wheel (statistical moments of all the harmonics of the regular, residual and difference parts etc...) were "normalized" with respect to the population of the same type of machine (spur/helical) at the same speed (by normalization we mean calculation of the Mahalanobis distance). A matrix of these non-dimensional features of a gear wheel was displayed and the exceptional recordings were selected (see Figure 7).

The discrimination between the different failure modes was based on automatic screening of features and visual inspection of the phase averages of the first harmonics and the sum of harmonics.

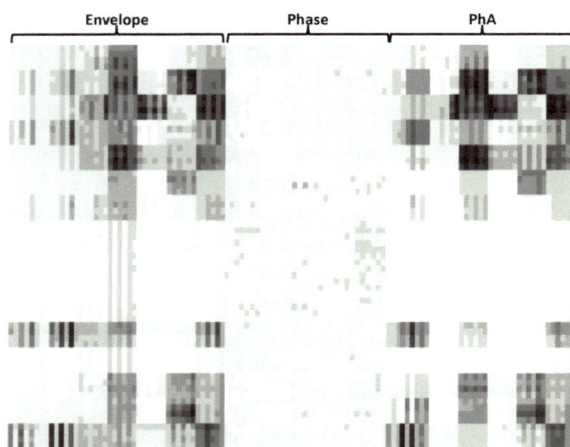

Figure 7. Scores of output spur gear – recordings at 50 RPS. The left third represents harmonics 1-5 (and over-all) scores of the envelope. The middle third represents the phase scores of harmonics 1-5 (and over-all), and the right third represents the phase-average scores of harmonics 1-5 and over-all. The scores of the idler gearwheel and output gearwheel are alternating.

The distributed faults (error related to the teeth spacing) were revealed by the RMS of the first harmony in all the types of phase averaged signals. The high RMS levels were most emphasized in the envelope of the first harmonic. In the orders domain appearance of sidebands was observed mainly around the first harmonic.

The localized faults were emphasized in the higher harmonics in all the statistical moments and in all the parts of the phase average. In the orders domain numerous high orders sidebands were elevated.

The fault location in the case of one of the wheels on the idler shaft was decided based on the orders representation, the comparison of the sensors, and the effect on the meshing gears on the other shafts.

It should be pointed out that all the signatures and hence features described are affected by mechanisms associated with the cross-gear interference, i.e. the vibrations induced by one gear (gear meshing) are modulated by the vibrations of another gear (gear meshing). For instance, in the PHM apparatus, a fault in the input pinion gear may cause a modulation of the pinion tooth meshing frequency with the tooth meshing of the large gearwheel, resulting in an erroneous identification of the fault location. Demodulation is impractical in the PHM apparatus since the gear mesh orders are exact multiples.

3.4 An example of a Distributed Fault Diagnostics

The diagnostic process of a distributed fault is illustrated using results from a recording classified as a spur gear at 50 RPS.

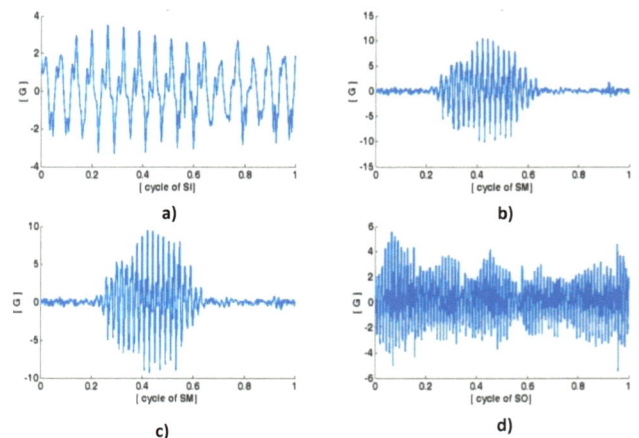

Figure 8. Phase averages of spur gear wheels from one recording with a faulty gear: a) input gearwheel, b) input idler gearwheel, c) output idler gearwheel, and d) output gearwheel.

The phase average of all the gear wheels is presented in Figure 8. Figure 9 presents the first harmonic of the phase average of the gear wheels. In both figures the fault can be easily observed on both wheels on the idler shaft (SM) while the amplitude is higher on the input gear wheel.

In Figure 9 it can be seen that the phase average signal is less noisy on the output gearwheel of the idler shaft than on the input gearwheel of the same shaft.

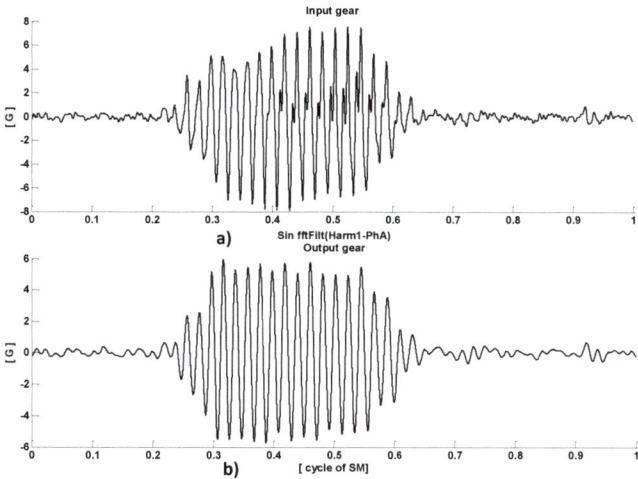

Figure 9. Phase averages of the first harmonic of spur gearwheels on the idler shaft. This example was taken from a recording with a faulty gear. a) Input gearwheel and b) output gearwheel.

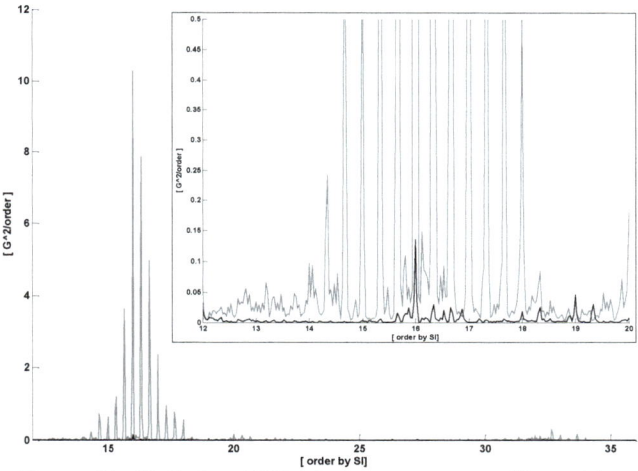

Figure 10. SI-Orders PSD of a spur gear recording with a faulty gear (grey) compared to a healthy gear (black)

In order to identify the fault location we checked several criteria:

• The influence on the meshing gear wheels: If the faulty gear wheel is on the input gear we would expect a higher impact on the gear wheel of the input shaft and the expected effect is a "smeared" envelope similar to the envelope on the faulty gear wheel. If the faulty wheel is on the output gear we would expect a signal with an envelope similar to the faulty gearwheel repeated 5 times. This can be observed in Figure 8.

• The FM modulation represented in the orders domain: FM modulation is expected to appear more clearly around the gearmesh of a faulty gear. In the case of spur gears, if the faulty wheel is on the input gear we would expect to see it around SI-order 32, and in the case that the fault is on the wheel of the output gear, around SI-order 16. In our case the fault is clearly located on the output gearwheel (see Figure 10).

Inspecting the statistical moments of the phase average, residual, difference, and their complex envelopes (Figure 7) we can see that the faults are manifested especially in the second statistical moments of the envelope. Of these the most relevant are the moments of the first harmonic and the overall signals (1-5 harmonics, see Figure 11).

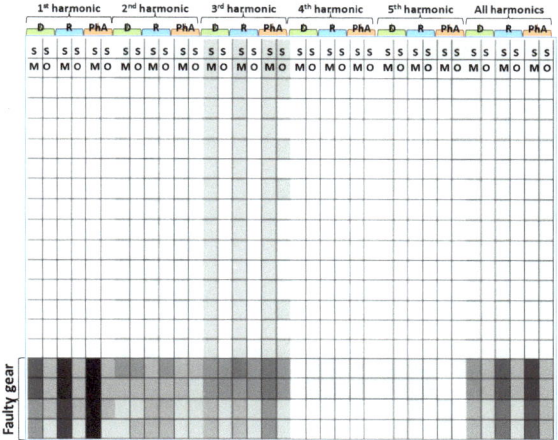

Figure 11. Second order statistical moments of the envelopes of: phase average (PhA), residual (R), and difference (D); harmonics 1-5 and over-all of alternating idler and output gearwheels. The lower four rows represent faulty idler gearwheels.

4 SHAFT ANALYSIS

The orders representation of the phase average according to each shaft is the most significant signature in the shaft analysis.

4.1 Shaft features

The features extracted for the shafts were the peak values of the first 5 harmonics of the respective phase average signals.

4.2 Shaft diagnostics

Unbalance of the input shaft was diagnosed based on the shaft rotating speed (1st order).

The probability for bent shaft was estimated based on the energy of the first few harmonies of the shaft rotating speed.

5 CONCLUSIONS

- Spectra in the frequency and orders domains did not always reveal the bearing faults. However, failures of the bearings are detectable by either envelope of the de-phased signal and envelope of the signal filtered around resonance frequencies.

- Fault localization on similar bearings on the same shaft (in the PHM challenge case, discrimination between the input and output side bearings on a shaft) is problematic without marked cases of healthy and faulty bearings.

- Detection of bearing faults requires high resolution (which for a given bandwidth can be obtained with longer recording time).

- Detection of bearing faults requires wide-band data regardless of the fact that commonly the pointers are at low frequencies (the resonance frequencies enhancing the bearing tones are much higher than the rate of shocks representing the contact with the damaged surface).

- De-phase process performs well under steady-state conditions and allows enhanced bearing detection even if the number of averages is low.

- The diagnostics of bearings can be achieved using fully automatic algorithms.

- Distributed gear failures are detectable using statistical moments of the envelope of the phase average, residual and difference signals (especially around first gearmesh harmony).

- Correct classification of operation conditions can greatly improve reliability of diagnostics. In the PHM case, operation conditions were uncertain due to unmarked load variations (high, low, bad key).

- Automatic localization of the faulty gear wheel is challenging due to the fact that wheels on the same shaft are indistinguishable. Cross-gear interference may lead to erroneous identification of faults.

NOMENCLATURE

$e(t)$	envelope of the signal
FM	Frequency Modulation
$H\{\}$	Hilbert transform
PhA	Phase Average
P_{xx}	Welch's PSD estimate
R	residual signal
D	difference signal
PSD	Power Spectral Density
RMS	Root Mean Square
RPS	Rotations Per Second [Hz]
SI	input shaft rotating speed
SM	idler shaft rotating speed
SO	output shaft rotating speed
$x(t)$	signal

REFERENCES

Antoni J. & Randall R. B. (2002), Differential Diagnosis of Gear and Bearing Faults, *Journal of Vibration and Acoustics*, Vol. 124, pp165-171.

Azovtsev A. & Barkov A. (1984), Rolling Element and Fluid Film Bearing Diagnostics Using Enveloping Methods, *Vibro Acoustical Systems and Technologies, Inc.*

Klein R., Rudyk E., Masad E. & Issacharoff M. (2009), Emphasizing bearings' tones for prognostics, *The Sixth International Conference on Condition Monitoring and Machinery Failure Prevention Technologies*, pp. 578-587.

Lebold M., McClintic K., Campbell R., Byington C. & Maynard K. (2000), Review of vibration analysis methods for gearbox diagnostics and prognostics, *Proceedings of 54th Meeting of the Society for Machinery Failure Prevention Technology*, Virginia Beach, VA, May 1-4, 2000, pp. 623-634.

McFadden P. D. & Smith J. D. (1984, February), Vibration Monitoring of rolling element bearing by the high-frequency resonance technique - a review, *Tribology international*, Vol 17, pp 3-10.

McFadden P. D. (1990), Condition monitoring of rolling element bearing by vibration analysis, *Proceedings of the Inst. Mech. Eng., Seminar on machine condition monitoring*, pp 49-53, 1990.

Sawalhi N. & Randall R. B., Semi-automated bearing diagnostics – three case studies, *School of Mechanical and Manufacturing Engineering, The University of New South Wales*, Sydney, Australia.

Wang Y. F. & Kootsookos P. J. (1998), Modeling of Low Shaft Speed Bearing Faults for Condition Monitoring, *Mechanical Systems and Signal Processing*, 12(3) pp 415-426.

Yang W. X. & Ren X. M. (2004), Detecting Impulses in Mechanical Signals by Wavelets, *EURASIP journal on Applied Signal Processing*, pp 1156-1162.

Zacksenhouse M., Braun S., Feldman M. & Sidahmed M. (2000), Toward Helicopter Gearbox Diagnostics from a Small Number of Examples, *Mechanical Systems and Signal Processing*, 14(4) pp 523-543.

Renata Klein received her B.Sc. in Physics and Ph.D. in the field of Signal Processing from the Technion, Israel Institute of Technology. In the first 17 years of her professional career she worked in ADA-Rafael, the Israeli Armament Development Authority, where she managed the Vibration Analysis department. In the decade that followed, she focused on development of vibration based health management systems for machinery. She invented and managed the development of vibration based diagnostics and prognostics systems that are used successfully in combat helicopters of the Israeli Air Force, in UAV's and in jet engines. Renata was a lecturer in the faculty of Aerospace Engineering of the Technion, where she developed and conducted a graduate class in the field of machinery diagnostics. In the last three years, Renata is the CEO and owner of "R.K. Diagnostics", providing R&D services and algorithms to companies who wish to integrate Machinery health management and prognostics capabilities in their products.

Eduard Rudyk holds a B.Sc. in Electrical Engineering from Ben-Gurion University, Israel, M.Sc. in Electrical Engineering and MBA from Technion, Israel Institute of Technology. His professional career progressed through a series of professional and managerial positions, leading development of pattern recognition algorithms for medical

diagnostics and leading development of health management and prognostics algorithms for airborne platforms, such as UAV's and helicopters. For the last 2 years Eduard is the director of R&D at "R.K. Diagnostics".

Eyal Masad received his B.Sc., M.Sc. and Ph.D. degrees from the Faculty of Mathematics in the Technion, Israel Institute of Technology. His research topics were in the fields of Machine learning, Information theory, nonlinear analysis and topological dynamics. In the last two years, Eyal is an algorithms developer at "R.K. Diagnostics".

Moshe Issacharoff holds a B.Sc. in Electrical Engineering from the Technion, Israel Institute of Technology. Moshe started his professional career in ADA-Rafael, the Israeli Armament Development Authority. During the last 2 years Moshe is an algorithms developer at "R.K. Diagnostics".

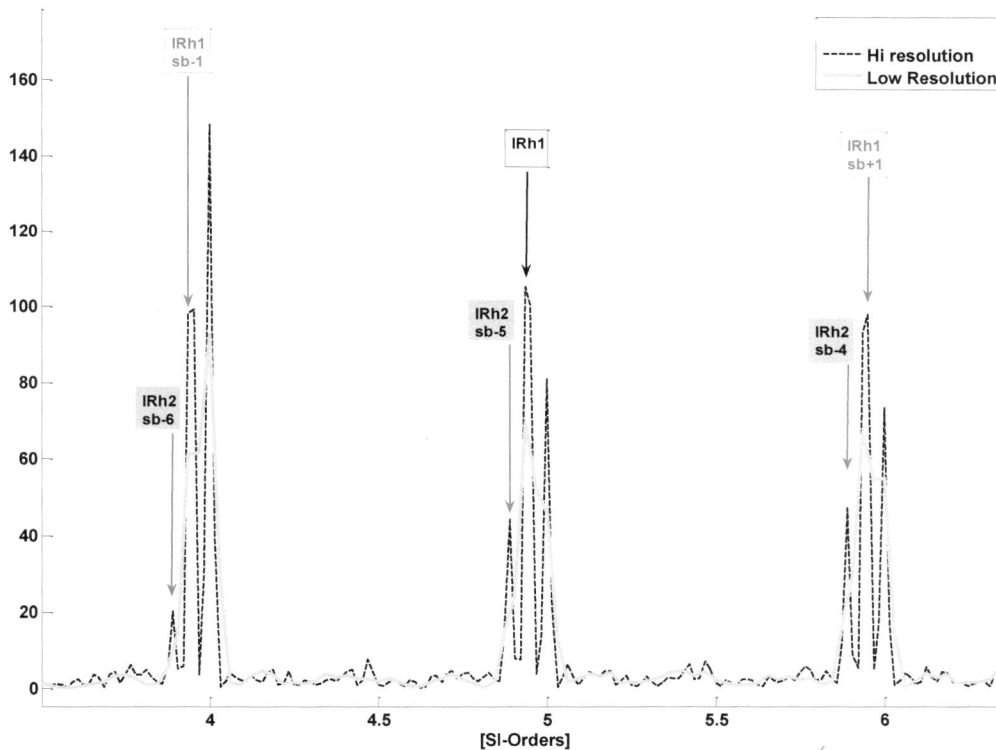

Figure 12. Comparison of low resolution (solid grey) and high resolution (dashed) of the orders by SI of the envelope of the de-phased signal. IR patterns of various harmonies are visible in the high resolution (h stands for harmony and sb stands for sideband).

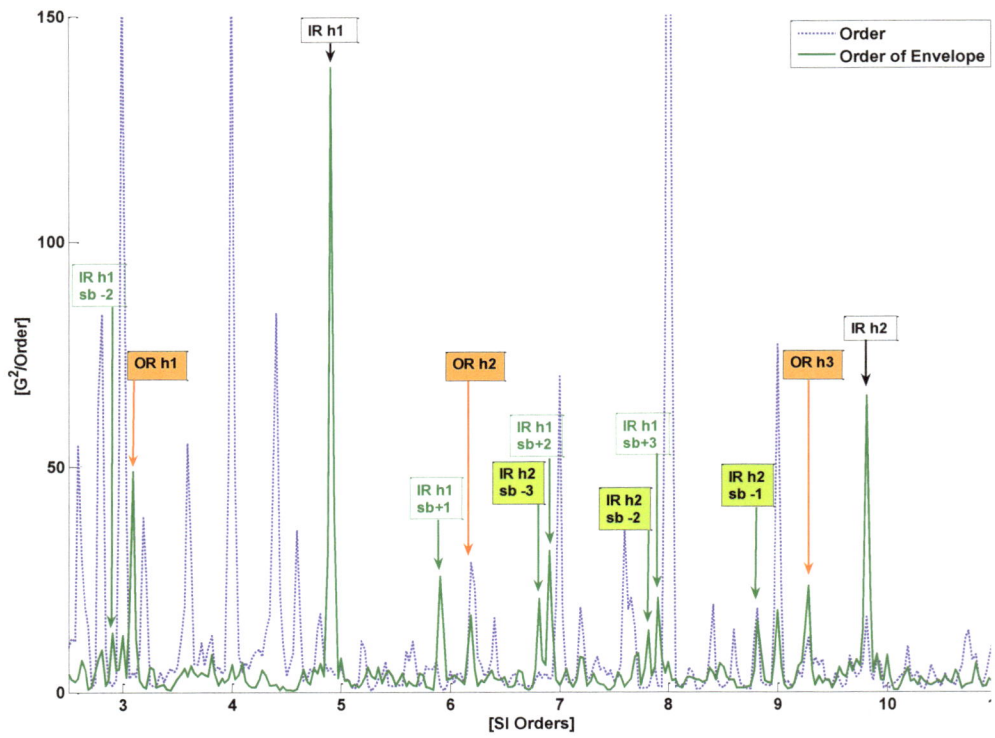

Figure 13. SI-order PSD (dotted) and an SI-order envelope PSD (solid). The arrows show some of the bearing pointers.

INTERNATIONAL JOURNAL OF PROGNOSTICS AND HEALTH MANAGEMENT, VOL.2 (2011)

Engine Oil Condition Monitoring Using High Temperature Integrated Ultrasonic Transducers

Kuo-Ting Wu[1], Makiko Kobayashi[1], Zhigang Sun[1]*, Cheng-Kuei Jen[1], Pierre Sammut[1], Jeff Bird[2], Brian Galeote[2] and Nezih Mrad[3]

[1] *Industrial Materials Institute, National Research Council Canada, Boucherville, Quebec, Canada J4B 6Y4*
**zhigang.sun@imi.cnrc-nrc.gc.ca*

[2] *Institute for Aerospace Research, National Research Council Canada, Ottawa, Ontario, Canada K1A 0R6*

[3] *Department of National Defence, Air Vehicles Research Section, Ottawa, Ontario, Canada K1A 0K*

ABSTRACT

The present work contains two parts. In the first part, high temperature integrated ultrasonic transducers (IUTs) made of thick piezoelectric composite films, were coated directly onto lubricant oil supply and sump lines of a modified CF700 turbojet engine. These piezoelectric films were fabricated using a sol-gel spray technology. By operating these IUTs in transmission mode, the amplitude and velocity of transmitted ultrasonic waves across the flow channel of the lubricant oil in supply and sump lines were measured during engine operation. Results have shown that the amplitude of the ultrasonic waves is sensitive to the presence of air bubbles in the oil and that the ultrasound velocity is linearly dependent on oil temperature. In the second part of the work, the sensitivity of ultrasound to engine lubricant oil degradation was investigated by using an ultrasonically equipped and thermally-controlled laboratory test cell and lubricant oils of different grades. The results have shown that at a given temperature, ultrasound velocity decreases with a decrease in oil viscosity. Based on the results obtained in both parts of the study, ultrasound velocity measurement is proposed for monitoring oil degradation and transient oil temperature variation, whereas ultrasound amplitude measurement is proposed for monitoring air bubble content.

1. INTRODUCTION

Engine oil condition includes oil viscosity, oil cleanliness, air bubble content, and oil temperature. Being able to monitor these parameters will not only safeguard engine operation under an oil condition it was designed for, but will also provide a means to assess the health of the entire engine system since any deviation from the nominal state of these parameters could be linked to one or more faulty components. Among oil condition monitoring systems presently used in aircraft engines, inductive oil debris monitor sensor offered by GasTOPS can be fitted to oil lube line and is capable of counting and sizing ferrous and non-ferrous particles above minimum size; Quantitative Debris Monitor (QDM®) technology offered by Eaton Aerospace captures ferrous wear debris and counts ferromagnetic particles exceeding a pre-set mass threshold; Zapper® pulsed electric chip detector system, also offered by Eaton Aerospace, captures ferrous debris and issues a warning signal to prompt for an action when debris buildup bridges the gap between two electrodes. Ultrasound, when applied judiciously and under favorable conditions, has the capability to sense all the aforementioned oil condition parameters. Although methods employing piezoelectric ultrasonic transducers (UTs) have been widely used for real-time, in-situ or off-line non-destructive evaluation (NDE) of large metallic structures including airplanes, automobiles, ships, pressure vessels, pipelines, etc., owing to their subsurface inspection capability, fast inspection speed, simplicity and cost-effectiveness (Gandhi et al., 1992; Ihn et al., 2004; Birks et al., 1991), applications of piezoelectric UTs to engine condition monitoring are relatively few due to difficulties in implementing UTs at elevated engine operating temperatures. In the present work, integrated UT (IUT) and associated wiring assembly designed for engine condition monitoring were fabricated directly onto the lubricant oil supply and sump lines of a modified CF700 turbojet engine. The applicability of the IUT assemblies to real-time engine condition monitoring was then investigated. The engine conditions of interest were air bubble content in the oil supply line, oil viscosity degradation and temperature. In order to assess the capability of ultrasound for oil viscosity degradation monitoring, four lubricant oils were tested in a temperature range of 50 °C to 130 °C by using an ultrasonically equipped test cell.

In the present paper, actual engine and laboratory tests setups and results are presented. Based on the results, ultrasonic approaches for real-time monitoring of air bubble content in oil, oil viscosity degradation, and oil temperature are proposed. Some implementation details and advantages associated with the proposed approaches are discussed.

2. ENGINE OIL CONDITION MONITORING EXPERIMENT

A modified CF700 turbojet engine (with fan module removed), shown in Figure 1, was used as representative of typical vibration, temperature and oil environments, as well as component materials and surface finishes. This engine platform offered the opportunity to assess oil flow in typical fully developed supply line flows and substantially aerated sump return line flows.

Figure 1. A modified CF700 turbojet engine.

The fabrication of IUTs involves a sol-gel based sensor fabrication process (Kobayashi et al., 2004, 2009) consisting of six main steps: (1) preparation of high dielectric constant lead-zirconate-titanate (PZT) solution, (2) ball milling of piezoelectric bismuth titanate (BIT) or PZT powders to submicron size, (3) sensor spraying using slurries from steps (1) and (2) to form a thin UT film, (4) heat treating to produce a thin solid BIT composite (BIT-c) or PZT composite (PZT-c) film, (5) Corona poling to obtain piezoelectricity, and (6) electrode painting for electrical connectivity. Steps (3) and (4) are repeated multiple times to produce optimal film thickness for specified ultrasonic operating frequency and performance. Silver or platinum paste was used to fabricate top electrodes. BIT-c film is to be used for higher temperature applications (e.g. temperature endurance of 500 °C) (Kobayashi et al., 2004, 2009); whereas PZT-c film is to be used when higher piezoelectricity than BIT-c film is required and when temperature is lower than 200 °C. A series of aging and thermal shock resistance tests have been conducted on sprayed-on films. In one test, a PZT-c film was subjected to 375 thermal cycles between 22 °C and 150 °C. In one aging test, a PZT-c film was heated in a furnace at 200 °C for over 300 days. In another aging test, a PZT-c film was excited with a 125 Volt pulse at 1 kHz pulse repetition rate for 1500 hours. The films performed well in all the tests.

Sump Line Pipe
(O.D. = 15.60 mm, wall thickness = 0.80 mm)

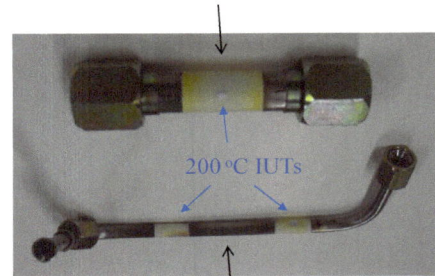

200 °C IUTs

Supply Line Pipe
(O.D. = 7.92 mm, wall thickness = 1.09 mm)

(a)

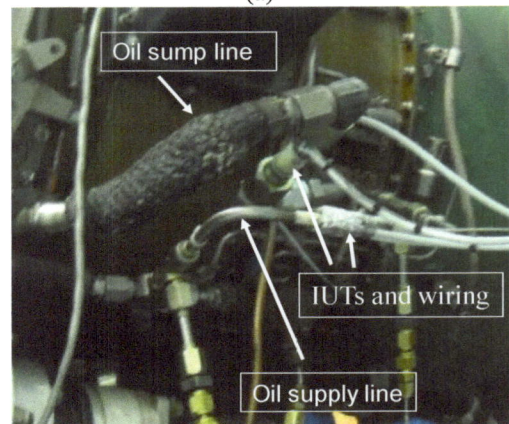

Oil sump line

IUTs and wiring

Oil supply line

(b)

Figure 2. PZT-c film IUTs (200°C) coated onto the sump and supply lines pipes before installation on the engine (a) and installation of ultrasonically instrumented oil sump and supply lines (b).

The present work aimed at performing monitoring during an actual engine test. PZT-c film IUTs were first coated on metal tubing of sump and supply lines (Figure 2(a)) and then these tubes were reinstalled in the engine as shown in Figure 2(b). The cables used could sustain temperatures of up to 200°C. The center frequencies of these PZT-c film IUTs were in the range of 10 to 12 MHz. The IUTs were operated in transmission mode whereby one IUT was used as a transmitter and another on the opposite side of the tube as a receiver. Two ultrasonic systems were used to monitor simultaneously oil in the supply and sump lines. Ultrasonic diagnostic signals were generated at a pulse repetition rate of 10 Hz, and digitized at a sampling rate of 100 MHz. The digitized signals were sent to a remote computer via a 40 meter-range USB communication adaptor for processing. In the meantime, oil temperatures in the sump and supply lines as well as engine speed, together with other engine operation related parameters were measured and recorded with a separate data acquisition system (DAS). The oil used was grade BPTO 2380 by Air BP USA.

Table 1 describes a test schedule developed as representative of transient and steady state operations. Data recording by DAS was controlled manually. Both steady state (SS) and transient (TR) recordings of engine data were conducted at every 50 milliseconds (20 Hz) when the engine speed was changed, after stabilization of two minutes, and after engine shutdown. The points and transitions selected are representative of basic performance tests in a test cell. The maximum speed is limited to 80% because of the matching of the engine and test cell airflow capabilities.

Sequence	Case	Notes	Status to record (SS: steady state TR: transient state)
1	Start-up	Take point before startup and during transient to idle	SS, TR
2	Idle	5 min @ idle, 50%, take points @ 2 min and 5 min	SS,SS
3	Slow accel.	30 second acceleration to 65%, hold for 2 min	TR, SS
4	Slow accel.	30 second acceleration to 80%	TR
5	Hold @ 80%	5 min hold @ 80%	SS
6	Slow decel.	30 second deceleration to 65%, hold, take points @ 2 min	TR, SS
7	Slow decel.	30 second deceleration to idle, hold, take points @ 5 min	TR, SS
8	Slow accel.	30 second acceleration to 65%, hold for 2 min	TR, SS
9	Slow accel.	30 second acceleration to 80%	TR
10	Hold @ 80%	5 min hold @ 80%	SS
11	Slow decel.	30 second deceleration to 65%, hold, take points @ 2 min	TR, SS
12	Slow decel.	30 second deceleration to idle, hold, take points @ 2 min	TR, SS
13	Shut-down	shut down, take SS after engine has stopped	TR, SS

Table 1 Schedule of an engine test

3. STATIC ULTRASONIC MEASUREMENT

A test cell shown in Figure 3 was used for studying the sensitivity of ultrasound to oil viscosity. The data will be presented in the Results and Discussion section. The ultrasound probes were composed of a stainless steel rod and a PZT-c film deposed on one end of the rod. The probes were intrusively mounted in the test cell with the inside end being in touch with the oil sample. One probe was used as transmitter and the other as receiver. A thermocouple was used for oil temperature measurement. Two oil samples of grade BPTO 2380 and two samples of a generic car motor oil with viscosity rating of SAE 5W20 were tested. Of the two BPTO 2380 samples, one was fresh and the other was degraded due to long term storage. These two samples can be easily differentiated from their colors as shown in Figure 4. For the two generic car motor oil samples, one was in fresh condition, and the other was used. Their difference could also be seen from their color difference. During the testing, the test cell was wrapped with a piece of thermal insulation fabric and was heated up to a pre-set temperature (130 °C). After the test cell reached thermal equilibrium, the heater was turned off and then data acquisition was carried out when the test cell was cooled naturally in a 23 °C room temperature setting. The cooling rates were about 1.3 °C/minute and 0.3 °C/minute when the oil temperatures were at 112 °C and 40 °C, respectively. We estimated that these cooling rates were slow enough to ensure a uniform temperature distribution in the probed section. The distance between the probing ends of the ultrasound probes was determined at room temperature by using water as sound wave propagation medium in the test cell and by measuring the transit time, t, for ultrasonic waves to travel the distance between probes ends. This distance was measured to be $d=c \times t=9.59$ mm, where c is the sound speed in water at 23 °C.

In a parallel experiment, the viscosities of the four oil samples were measured using a constant stress rotational rheometer, SR-200, presented in Figure 5. The strain response as a function of time under a constant stress load was monitored in a Couette type testing environment, at room temperature. The measured oil viscosities, resulting from the step stress (creep) measurements, are listed in Table 2.

BPTO 2380		Generic 5W20	
Fresh	0.0448	Fresh	0.1041
Old	0.0406	Used	0.0841

Table 2 Oil viscosity at 22 °C (Pa·s)

(a)

(b)

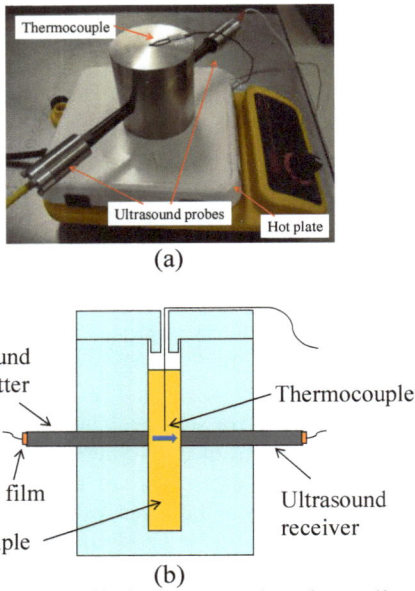

Figure 3. Ultrasonically instrumented testing cell used in the study. (a) picture of the setup; (b) schematic of a section view.

Figure 4. Fresh (left) and degraded (right) BPTO 2380 oil samples.

Figure 5. Rotational rheometer (Rheometrics SR-200) used for oil viscosity measurement.

4. RESULTS AND DISCUSSION

Figure 6 displays the variation of engine speed during the test described in Table 1. The asterisks represent the average of 20 samples acquired within a one-second period when the

engine stabilized whereas the solid lines represent the speed when the engine was in a transient state.

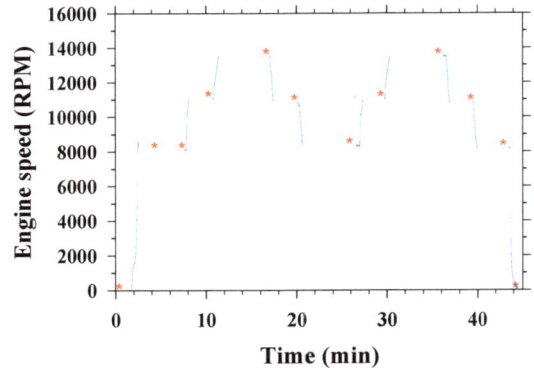

Figure 6. Variation of engine speed during the test scheduled in Table 1. The asterisks represent the steady state recording and the solid lines represent the transit state recording.

Figure 7 illustrates variation of oil temperature, measured with a T-type thermocouple, during the same engine test. By comparing Figures 6 and 7 one can see that oil temperature increased with increase of engine speed. The oil in the supply line was at higher temperatures than that in the sump line.

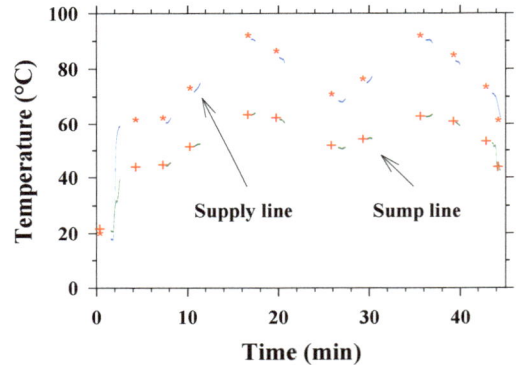

Figure 7. Variations of engine oil temperatures in the supply and sump lines. The asterisks represent the steady state recording and the solid lines represent the transit state recording.

Figure 8 displays one ultrasonic signal acquired in the supply line (a) and one in the sump line (b). In these recordings, not only the first transmitted signal (1st arrival), but also the second arrival resulting from one round-trip reflection of the first transmitted signal inside the oil flow channel are seen clearly. Since the sump line tube has a larger inner diameter, the ultrasonic signals across it arrive later than those in the supply line. Knowing the inner diameter of a pipe, d, and time delay between the 1st and 2nd signal arrivals, τ, the sound velocity in the oil can be obtained as $V=2d/\tau$.

Figure 9 displays the variation of the signal-to-noise ratio (SNR) of the 2nd arrival of the transmitted signals shown in

Figure 8. The signal-to-noise is defined as 20log10(Asig/Anoise) with Asig and Anoise being respectively the peak-to-peak amplitude of the signal and the peak-to-peak amplitude of the noise measured just before the signal arrival. The display covers about 40 minutes of process time with each data point being the SNR of one acquired signal. Since ultrasound signals were generated at a pulse repetition rate of 10 Hz, about 24,000 data points are displayed for each test result. A large signal-to-noise ratio means a strong transmitted signal. While the SNR of signals measured in the supply line remains high and quasi constant, the SNR of signals measured in the sump line fluctuates significantly. For two reasons the signal quality fluctuation in the sump line is believed to be caused by the presence of air bubbles in the oil. First, the high sensitivity of ultrasound amplitude to air bubbles has been proven by extensive experimental and theoretical work (Leighton T., 1996). Second, air bubbles do get trapped in the oil circuit. The air was forced into the oil mainly by the high speed rotation of the bearing (and the parts in it like the balls and cage). The foamy oil left the bearing, passed into a sump and was drawn through the sump line to an oil/air separator. The recovered oil was then sent back through the supply line to the oil circulating system. The variation of SNR indicates that the quantity of air in the oil varied during the test. At the beginning of the test, the oil in the sump line had few bubbles. This translates into strong signals (high SNR).

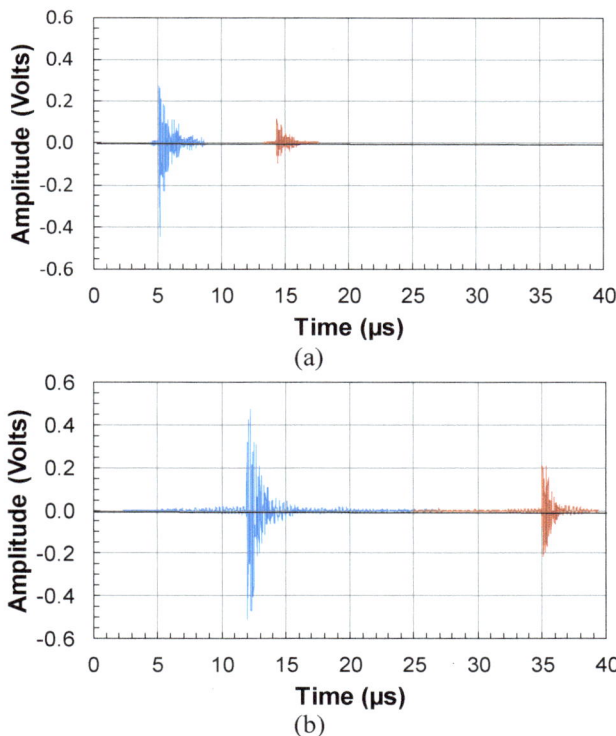

(a)

(b)

Figure 8. A trace of ultrasound signal recorded with supply line IUTs (a) and sump line IUTs (b).

On comparing Figures 6 and 9, overall, a higher engine speed with higher oil flow rate and temperature appears to lead to larger counts of low SNR. The signal quality in the supply line was stable because the oil was pumped and filtered before being sent to the ultrasonically probed section of the supply line. The sensitivity of ultrasound to air bubbles indicates that ultrasound can be used to evaluate the quantity of air bubbles in the engine oil supply system. Thus condition monitoring assessments might be made of the oil system integrity and oil condition. A preliminary study on using this technique for detection of metal particles was carried out previously (Kobayashi et al., 2007). Further studies with injected debris are underway to assess if the approach may also be used to detect the presence of metal debris in the oil circuit.

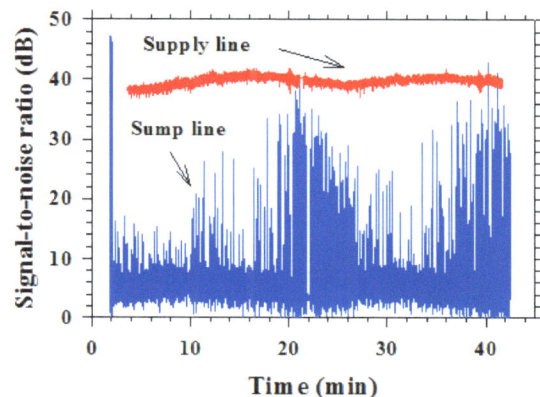

Figure 9. Variation of the signal-to-noise ratio of the 2nd arrival of transmitted signals in the supply (upper curve) and sump (lower curve) lines.

Figure 10 illustrates the variation of ultrasound velocity in the engine oil in the supply line (solid line). The circled cross hairs represent the ultrasonic measurement data recorded at the same times the steady state recordings, shown in Figures 6 and 7, were taken by the DAS. Figure 11 shows the variation of ultrasound velocity versus temperature using the steady state recorded data. Also shown in the figure is a linear regression fit of the sound velocity versus oil temperature data, illustrating the linear dependence of ultrasound velocity on engine oil temperature.

The measured viscosities of the oil samples (BPTO 2380 and generic 5W20) are listed in Table 2. Figure 12 presents the relationship between ultrasound velocity and temperature for the four oil samples measured with the test cell illustrated in Figure 3. Significant difference is observed between oil families BPTO 2380 and generic 5W20. In order to better assess the sensitivity of ultrasound velocity to the freshness state of the oil, closer views are provided in Figures 13 and 14. For each oil sample, several tests were performed to evaluate the consistency of the results. For each oil family, the fresh sample, which possesses a higher viscosity, has a higher ultrasound velocity. This indicates that oil viscosity degradation can be assessed by monitoring the decrease in ultrasound velocity. However, the ultrasound velocity is quite sensitive to oil temperature. The ultrasound velocity difference between the fresh and the used (or degraded) oil samples can be easily offset by a temperature variation of merely 0.7 °C. This makes in-flight ultrasonic oil viscosity monitoring more challenging since the oil temperature has to be accurately monitored to compensate for ultrasound velocity change by temperature. This difficulty could be overcome if the following conditions are observed: (1) The thermocouple used for oil temperature measurement performs consistently over time; (2) a fairly large number (e.g. 100) of ultrasound velocity and oil temperature readings are taken in a short period of time during which the oil temperature change is considered negligible. Then the averages of these readings (which are much less affected by random noise in ultrasound and temperature signals) can be used to represent the physical state of the oil. Practically speaking, today's thermocouples can be made to be inexpensive and robust. Not only does this mean that long term reliability and consistency of thermocouples are generally guaranteed, but it also means that more than one thermocouple could be implemented economically to cross-check oil temperature, thus significantly reducing the chance of having erroneous readings. The second requirement can also be fulfilled easily given that today's data acquisition systems can deliver hundreds of readings per second with ease and that stable oil temperature is achievable after the engine enters a steady state. Although technically achievable, the necessity of precise oil temperature measurement, usually through an intrusive temperature sensor, would negate the benefit of non-invasive ultrasonic viscosity measurement. A pure acoustic viscosity measurement solution is highly desirable and will be the subject of a follow-up research.

It is observed in Figures 13 and 14 that the sound velocity variation induced by a change in oil viscosity would not exceed that caused by 1 °C of oil temperature change. This indicates that ultrasound velocity can be used for oil temperature sensing with guaranteed uncertainty smaller than 1 °C regardless of oil freshness state. Compared with conventional thermocouples, ultrasonic temperature measurement has the advantages of having instant response, providing average temperature over the entire monitored section instead of a limited area at the sensing tip of a thermocouple, and not being affected by the temperature of the oil tube on which the ultrasound transducers are mounted. These beneficial features may make ultrasonic temperature measurement particularly attractive for capturing abnormally fast oil temperature surges which may come as an early sign of a deficient engine component.

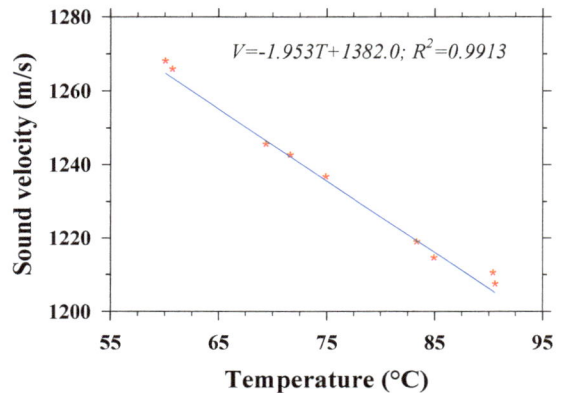

$$V = -1.953T + 1382.0; \ R^2 = 0.9913$$

Figure 11. Variation of ultrasound velocity versus the temperature of engine oil in the supply line. The asterisks represent the steady state recording and the solid line represents a linear regression fitting.

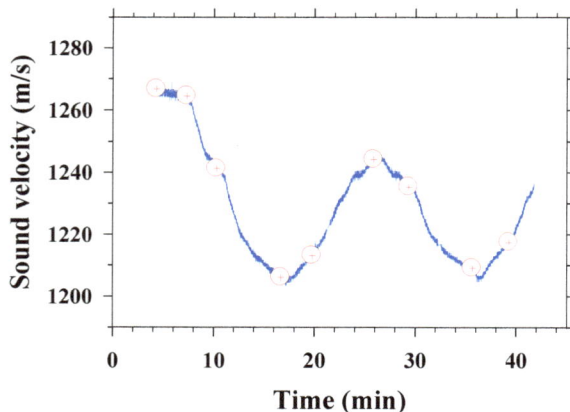

Figure 10. Variation of ultrasound velocity in the engine oil in the supply line (solid line). The circled cross hairs represent the ultrasonic measurement data recorded at the same times the steady state recordings were taken by DAS.

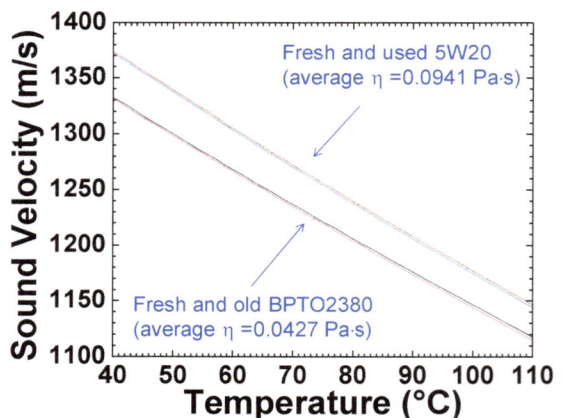

Figure 12. Ultrasound velocity versus temperature relationship for four oil samples measured with the testing cell shown in Figure 3.

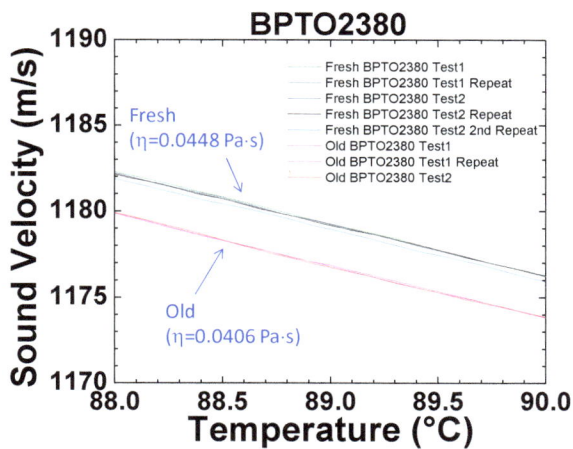

Figure 13. Ultrasound velocity versus temperature relationship for BPTO 2380 oils.

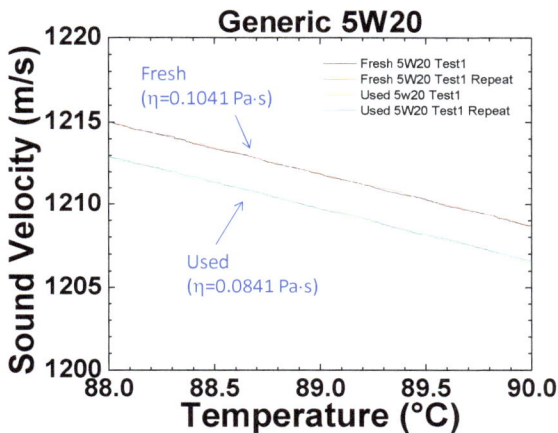

Figure 14. Ultrasound velocity versus temperature relationship for the generic 5W20 oils.

5. CONCLUSIONS

High temperature integrated ultrasound transducers made of thick piezoelectric composite films have been coated onto the lubricant oil supply and sump lines of a modified CF700 turbojet engine. These transducers have been used during an actual engine test cell operation to evaluate their potential usefulness in engine oil condition monitoring. Furthermore, a study on the sensitivity of ultrasound to engine lubricant oil viscosity degradation has been carried out using a thermally-controlled laboratory test cell and lubricant oils of different grades. The actual engine tests results and laboratory tests results have shown that the presented ultrasound technique can provide an effective non-intrusive means for detection of air bubbles in the oil supply line (oil system component degradation/failure) and for instantaneous oil temperature measurement. The capability

of the technique for oil viscosity measurement is compromised by the necessity of simultaneous oil temperature measurement. The applicability of the technique to absolute temperature measurement and to the detection of metal debris in oil needs to be further investigated in the future.

ACKNOWLEDGMENT

Technical supports of Harold Hébert from the Industrial Materials Institute (IMI) and Marcel Fournier from the Institute of Aerospace Research (IAR) of the National Research Council Canada (NRCC) are greatly appreciated. This work was carried out under collaborative research agreements between Defence Research and Development Canada (DRDC), National Defence (DGAEPM) and NRCC.

REFERENCES

Birks, A.S., Green, R.E. Jr. and McIntire, P. ed., (1991). *Nondestructive Testing Handbook*, 2nd Edition, vol.7, Ultrasonic Testing, ASNT.

Gandhi, M.V. and Thompson, B.S., (1992). *Smart Materials and Structures*, Chapman & Hall, NY.

Ihn, J.-B. and Chang, F.-K., (2004). Ultrasonic non-destructive evaluation for structure health monitoring: built-in diagnostics for hot-spot monitoring in metallic and composite structures, Chapter 9 in *Ultrasonic Nondestructive Evaluation Engineering and Biological Material Characterization*, edited by Kundu T., CRC Press, NY.

Kobayashi, M. and Jen, C.-K., (2004). Piezoelectric thick bismuth titanate/PZT composite film transducers for smart NDE of metals, *Smart Materials and Structures*, vol.13, pp.951-956.

Kobayashi, M., Jen, C.-K., Bussiere, J.F. and Wu, K.-T., (2009). High temperature integrated and flexible ultrasonic transducers for non-destructive testing, *NDT&E Int.*, vol.42, pp.157-161.

Kobayashi, M., Jen, C.-K., Moisan, J.-F., Mrad, N. and Nguyen, S.B., (2007). Integrated ultrasonic transducers made by sol-gel spray technique for structural health monitoring, *Smart Materials and Structures*, vol.16, pp.317-322.

LeightonT., (1996). Comparison of the abilities of eight acoustic *techniques to detect and size a single bubble, Ultrasonics*, vol.34, pp.661–667.

An Auto-Associative Residual Processing and K-means Clustering Approach for Anemometer Health Assessment

David Siegel[1] and Jay Lee[1]

[1]NSF I/UCRC for Intelligent Maintenance Systems (IMS), University of Cincinnati, Ohio, 45220, United States of America

siegeldn@mail.uc.edu
jay.lee@uc.edu

ABSTRACT

This paper presents a health assessment methodology, as well as specific residual processing and figure of merit algorithms for anemometers in two different configurations. The methodology and algorithms are applied to data sets provided by the Prognostics and Health Management Society 2011 Data Challenge. The two configurations consist of the "paired" data set in which two anemometers are positioned at the same height, and the "shear" data set which includes an array of anemometers at different heights. Various wind speed statistics, wind direction, and ambient temperature information are provided, in which the objective is to classify the anemometer health status during a set of samples from a 5 day period. The proposed health assessment methodology consists of a set of data processing steps that include: data filtering and pre-processing, a residual or difference calculation, and a k-means clustering based figure of merit calculation. The residual processing for the paired data set was performed using a straightforward difference calculation, while the shear data set utilized an additional set of algorithm processing steps to calculate a weighted residual value for each anemometer. The residual processing algorithm for the shear data set used a set of auto-associative neural network models to learn the underlying correlation relationship between the anemometer sensors and to calculate a weighted residual value for each of the anemometer wind speed measurements. A figure of merit value based on the mean value of the smaller of the two clusters for the wind speed residual is used to determine the health status of each anemometer. Overall, the proposed methodology and algorithms show promise, in that the results from this approach resulted in the top score for the PHM 2011 Data Challenge Competition. Using different clustering algorithms or density estimation methods for the figure of merit calculation is being considered for future work.

1. INTRODUCTION

One of the fundamental requirements for data interpretation, model development, and system monitoring is the need to have properly working and calibrated sensory data (Venkatasubramanian, Rengaswamy, Yin, K., & Kavuri, 2003). Considering the importance of properly working sensors, there is considerable research in the area of sensor fault detection and diagnosis with a diverse set of applications ranging from automotive (Capriglione, Liguroi, Pianese, & Pietrosanto, 2003), aerospace (Patton, 1991), to nuclear power plants (Hines & Garvey, 2006). The wind energy in particular, is quite reliant on obtaining accurate sensor measurements of wind speed, since this ultimately is one of the inputs used to estimate the energy production for a given site (Petersen, Mortensen, Landberg, Hujstrup, & Frank, 1998). During feasibility studies of potential wind turbine sites, anemometers placed on meteorological towers are used to provide information on the long term wind speed characteristics. Historical wind speed data is one of the inputs provided to sophisticated meteorological models that provide an estimation of the energy production for a given site. Errors in the wind speed measurements can have significant effects on the estimated energy production which could affect the return on investment for a given site or whether the site is financed (Murakami, Mochida, & Kato, 2003).

Recent work in the area of anemometer fault detection includes the work by Kusiak, Zheng, and Zhang (2011), which propose a virtual sensor method using a multilayer perceptron neural network. This study also discusses the use of a wavelet de-noising method for data pre-processing and a control chart based on the residuals calculated from the predicted and measured wind speed. A more classical statistical approach was discussed in the work by Beltran, Llombart, & Guerrero (2009), in which a metric was derived from the difference in the 10 minute wind speed average data between two anemometers in close proximity. In addition to this prior work, a recent study by Clark, Clay, Goglia, Hoopes, Jacobs, and Smith (2009) was done to investigate the root cause of NRG #40 anemometers reading slower than the actual wind speed. The study discussed statistical metrics, signatures of anemometers with excess measurement error, calibration methods, and a physical explanation of the potential failure mode that was believed to be the cause of the sensor measurement error.

The paper is organized in the following manner: after the introduction, Section 2 describes the problem statement for the 2011 Prognostics and Health Management Society Data Challenge. This is followed by an overview of the algorithms used for the shear and paired data sets in Section 3. More detailed descriptions of the filtering and data normalization methods are described in Section 4. The residual processing method using auto-associative neural network models is presented in Section 5. Section 6 describes the use of k-means clustering and the figure of merit health value. Lastly, conclusions and future work are discussed in Section 7 and Section 8 respectively.

2. PROBLEM STATEMENT

The 2011 Prognostics and Health Management Society (PHM Society 2011) presented a data challenge problem dealing with this increasingly important topic of anemometer fault detection. Two different types of data sets titled the "paired data set" and the "shear data set" was provided for developing and evaluating anemometer fault detection algorithms. The paired data set consisted of data collected from two anemometers at the same height. Statistics from the two wind speed sensors, a wind direction measurement, and ambient temperature reading were provided. The statistics were calculated from a 10 minute time period and consisted of the mean, standard deviation, maximum, and minimum for each parameter. Data from paired anemometers in a nominal healthy condition were provided in 12 training data sets that comprised of 25 days worth of data. The competition also provided 420 test data files, in which each file contained 5 days worth of data. These test files were used to test the accuracy of the developed algorithm by the contest participants, in which the actual healthy state was unknown to the contest participants. In the test files, either one of the anemometers could be in a healthy or degraded state; the objective was to provide a correct healthy classification for both anemometers (PHM Society 2011).

The shear data set differed from the paired anemometer data set, in that there were either 3 or 4 anemometers and each anemometer was at a different height on the meteorological tower. Height information was provided for each anemometer and statistics from each anemometer were provided after processing the wind speed measurements in a 10 minute data block. As with the paired data set, the wind direction statistics and the ambient temperature statistics were also provided. In total, 28 or 23 parameters were provided in each data shear data file, the difference in the

number of parameters is due to certain sites only having 3 anemometers instead of 4 anemometers. A total of 7 training data sets that comprised of 25 days worth of data were provided for the shear data set; the training data sets provided data from anemometers in a healthy condition.

Test files were also provided in which the health condition of the anemometers were unknown to the contest participants. The test files consisted of 225 files, with each file representing 5 days worth of data. The objective in the shear data set was to determine whether the set of anemometers were all in a nominal healthy state or one or more of anemometers had a fault and were exhibiting excessive measurement error. Unlike the paired data set, it was not required to determine which anemometer was experiencing a fault. The requirements for the shear anemometer data set were to detect whether the system was in either a healthy or abnormal health state (PHM Society 2011).

3. ANEMOMETER HEALTH ASSESSMENT METHODOLOGY

The overall approach for assessing the health state of an anemometer consists of a series of algorithmic processing steps. These steps include data pre-processing, a residual calculation, and ultimately a decision on the health status based on a figure of merit metric. The health assessment algorithms developed for the shear and paired data sets follow that step by step processing methodology; however considering the unique aspects of both data sets, there are specific differences with regards to data normalization and the residual calculation. Section 3.1 and section 3.2 presents an overview of the methodology for assessing the health condition of the shear anemometers and paired anemometers respectively. More specific details of each processing module along with intermediate results from each step are shown in the subsequent sections to further illustrate the anemometer health assessment method.

3.1. Algorithm for Shear Data Set

A flow chart of the health assessment algorithm for the anemometer shear data is provided in Figure 1. The algorithm used in this study has a training and monitoring phase, in which the training phase is developed using shear anemometer data from a nominal healthy state. The initial step in the training process is to perform data filtering. The data filtering step is designed to remove instances in which icing could occur as well as to remove other data samples in which there could be erroneous readings in wind speed, temperature, or other sensor measurements. The data normalization step is a specific step designed for the shear data and is based on the wind profile power law (Peterson, & Hennessey Jr., 1977). This normalization procedure uses the power law equation to place each of the shear anemometer wind speed measurements at a common reference height.

The data normalization step reduces the variation due to elevation; however an auto-associative neural network is used to further model the relationship and correlation structure between the anemometer wind speed statistics. In this study, multiple baseline data sets were available for model training and this provided the opportunity to have multiple auto-associative neural network models. The training of the auto-associative neural network models completes the training process and the algorithm can then be deployed in its monitoring phase.

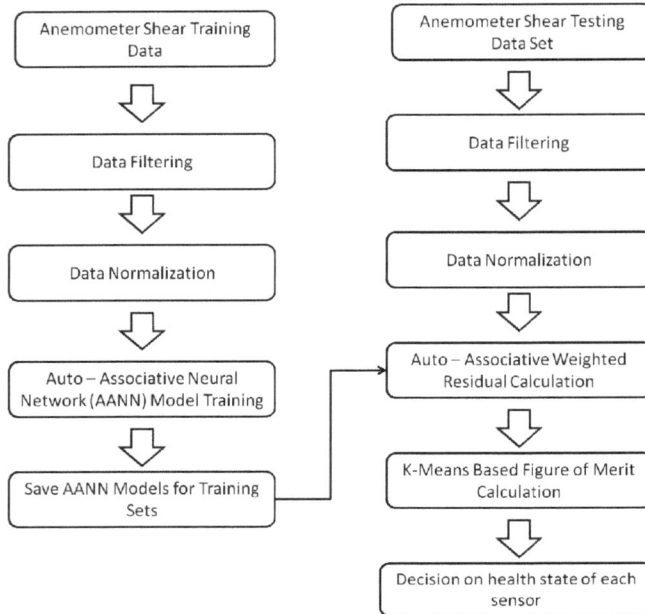

Figure 1. Algorithm Flow Chart for Shear Data Set

In the shear data set, a given monitored shear data set file consisted of either 3 or 4 anemometers and each data file comprised of 720 samples and a duration of 5 days. Thus, the data processing and health decision is performed on data from that 5 day period for a monitored shear anemometer set. The initial step for the monitored shear anemometers consist of performing the same data filtering and normalization that were used in the training set. A weighted residual calculation is performed using the auto-associative neural network models; a weighted approach is used in order to favor results from training models that more accurately predict the anemometer wind speed statistics. The residuals for the mean wind speed for each anemometer are then further processed in a k-means figure of merit calculation. The motivation for using a k-means clustering method is that prior literature suggested that the anemometers display a bimodal behavior in one of its failure modes and experience slowdown for a certain range of directions and wind speeds (Hale, Fusina, & Brower, 2011). Thus, the residuals might be quite small in a particular speed or direction regime and could be potentially quite higher in a different regime subset. A figure of merit calculation is

performed for each anemometer, and a decision on the health status for each anemometer is made on whether the figure of merit value exceeds the threshold.

3.2. Algorithm for Paired Data Set

The health assessment algorithm used for the paired data set can be considered a subset of the one used for the shear data set; in that the algorithm used for the paired data set does not require the additional data normalization or auto-associative neural network based residual processing. The flow chart in Figure 2 shows the processing steps for the anemometer paired data, in which the initial step includes a data filtering step to remove data instances when icing takes place as well as other erroneous samples. Considering that the paired data set consists of anemometers at the same height, it is not necessary to use the wind profile power law for normalizing the data to a common reference height.

Although it is conceivable to train an auto-associative neural network for the paired anemometers; the initial rationale was that this would be too complex of a modeling approach for this situation. A direct comparison between the wind speed mean values provides a simple but an effective way of inferring the health state of the wind speed sensor. The approach used in this study calculates the difference between the wind speed mean values, denoted as d_{12} and d_{21}, and uses the difference signals as a surrogate for the residual signal used in the shear data set. The same k-means figure of merit calculation used in the shear health assessment algorithm is than applied to the difference signal.

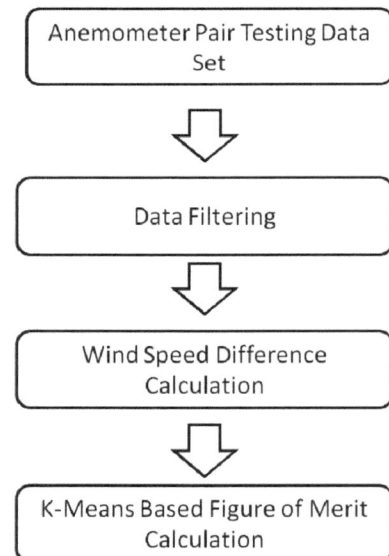

Figure 2. Algorithm Flow Chart for Paired Data Set

The primary focus of the health assessment algorithm for the paired data set was to provide a fault detection capability with the assumption that one of the anemometers was in a nominal healthy condition. However, during the course of a data file that represents a 5 day period, there potentially could be a cluster in which one anemometer was reading significantly lower than normal and another cluster where the other anemometer was reading significantly lower than normal. In this scenario, both figures of merit values could exceed the threshold and both anemometers would be reported as being in a degraded health state.

4. DATA PRE-PROCESSING

The subsequent processing steps in the health assessment algorithms for the shear and paired data set depend on quality data inputs. In both instances, data filtering is performed to remove erroneous data samples and provide a more suitable data set for further processing. An additional data normalization step is performed for the shear data set in order to place all wind speed measurements at a common reference height. Sections 4.1 and 4.2 provide the more specific details regarding the data pre-processing.

4.1. Data Filtering

The filtering routine is done to remove samples in which icing could be occurring and also for filtering out samples in which there are erroneous senor values. For removing instances in which icing is occurring, there is a variety of parameters that could be used to infer this condition; the wind speed direction standard deviation statistic in particular is quite useful for filtering out icing events. Considering that various statistics are calculated for each 10 minute data block, a value of zero in the wind speed direction standard deviation would imply that there is no variation in the wind speed direction for a 10 minute time period. Physically this is not possible and this condition of no variation in the wind speed direction is one of the key parameters that can be used for filtering out samples in which icing could occur.

The filtering settings used for the shear and paired data sets are provided in Table 1 and Table 2 respectively. For a given sample for the paired data set, it would have to satisfy all the listed ranges shown for the wind speed means, ambient temperature, wind direction mean, and wind direction standard deviation. It was observed in both the paired and shear training data sets that instances in which the wind direction were quite low resulted in more sudden changes in wind speed mean values. This resulted in larger differences between anemometer wind speed readings during these more abrupt changes. Considering this aspect, the filtering routine includes logic for the wind direction mean parameter to remove these samples in which the wind direction is below 50 degrees. It was also observed in the training data sets that the initial samples in each data file contained erroneous sensor values, thus the filtering routine also removed the first 20 samples.

The filtering routine for the shear and paired data set is quite similar, the major differences include that the paired data set filtering routine includes the anemometer wind speed standard deviation parameter. A low wind speed standard deviation would imply very little variation in the wind speed mean for a 10 minute period, which could imply icing. However, it was noted that including the anemometer wind speed standard deviation for the shear data set filter removed too many samples in a few of the test files, thus this setting was only used for the paired data set filtering routine.

Parameter	Filter Settings
Anemometer Mean 1	0.5 m/s – 26 m/s
Anemometer Mean 2	0.5 m/s – 26 m/s
Anemometer Mean 3	0.5 m/s – 26 m/s
Ambient Temperature Mean	-40 °C - 120 °C
Wind Direction Mean	Greater than 50 degrees
Wind Direction standard deviation	Greater than 0 degrees

Table 1. Filtering Settings for Shear Data Set

Parameter	Filter Settings
Anemometer Mean 1	0.5 m/s – 26 m/s
Anemometer Mean 2	0.5 m/s – 26 m/s
Anemometer 1 Standard Deviation	Greater than 1 m/s
Anemometer 2 Standard Deviation	Greater than 1 m/s
Ambient Temperature Mean	-40 °C - 120 °C
Wind Direction Mean	Greater than 50 degrees
Wind Direction standard deviation	Greater than 0 degrees

Table 2. Filtering Settings for Paired Data Set

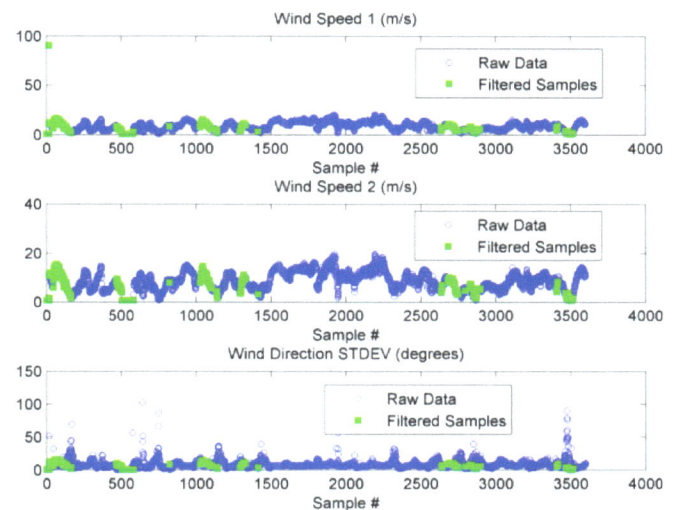

Figure 3. Raw and Filtered Out Samples – Training Set 6 for Shear Data Set

An example of how the filtering removes potentially icing events and erroneous data samples is shown in Figure 3. This example is from a shear training data set in which all 3 anemometer sensors are in a nominal healty condition; however, there is still a substantial amount of samples higlighted in green that are filtered out. The top plot highlights that the wind speed mean can have some extreme high or low values, as indicated by the outlier value near 100 and some of the values near or at zero. The middle plot shows the second anemometer wind speed mean reading and one can observe that there are several instances in which both anemometers are reading at or near zero. These near zero readings are likely due to icing. The wind direction standard deviation is shown in the bottom most plot and this parameter is also zero during these suspected icing samples. This example highlights that the filtering algorithm provides an adequate detection of icing and outlier samples.

4.2. Data Normalization

The data preprocessing for the shear data set includes an additional step of data normalization in order to compare the wind speed measurements at a common reference height. In prior work in the literature, the wind speed profile has been modeled as a logarithmic relationship and also by a power law model (Peterson et al, 1977). The use of the logarithmic equation includes an additional aerodynamic surface roughness parameter that depends on the site location; this was not provided in this study and thus only the power law equation was used for data normalization. The power law equation is described by Eq. (1), in which u_1 and z_1 are the wind speed and height at a known reference point and u_2 and z_2 are the wind speed and height at a location of interest. The exponent P is a constant that is based on prior experimental studies and regression fitting; a value of 1/7 is a common value for this constant and one that is used in this study (Hsu, Meindl, & Gilhousen, 1994).

$$\frac{u_2}{u_1} = \left(\frac{z_2}{z_1}\right)^P \tag{1}$$

For data normalization, each wind speed measurement is corrected to a height of 49 m. In Eq. (1), this would imply that z_2 is assigned a value of 49, while u_1 and z_1 are the known wind speed measurement and elevation for a given anemometer and u_2 is the corrected wind speed measurement at a height of 49 meters. An example of normalization process is illustrated in Figure 4 and Figure 5. Figure 4 is from the first shear training data set and is comparing the wind speed for anemometers 1 and 4. With regards to the numbering convention, anemometers 1-4 are sorted from the highest to the lowest height and in this example have a height of 59, 50, 30, and 10 meters respectively. As one can observe, there is significant

differences in the raw wind speed values for anemometer 1 and 4, these two anemometers have the largest difference in elevation. Figure 5 shows the normalized wind speed mean values for anemometer 1 and 4 from the same training data set. From visual observation, one can observe that the differences in the normalized wind speed values are lower when compared with the raw data.

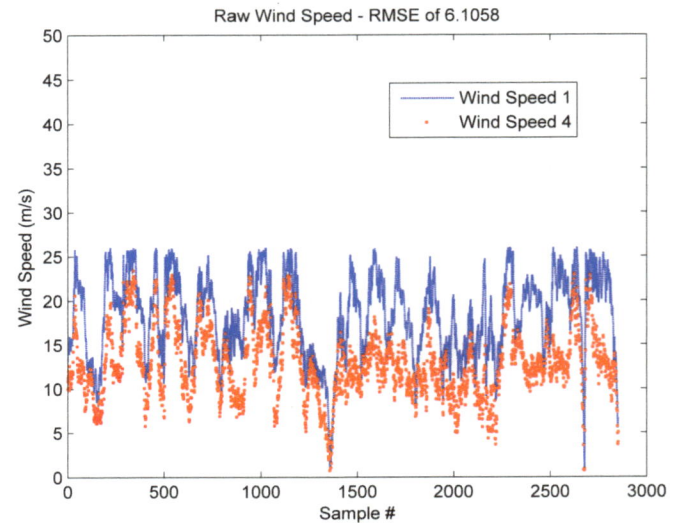

Figure 4. Shear Training Set 1 - Raw Wind Speed Mean Signals

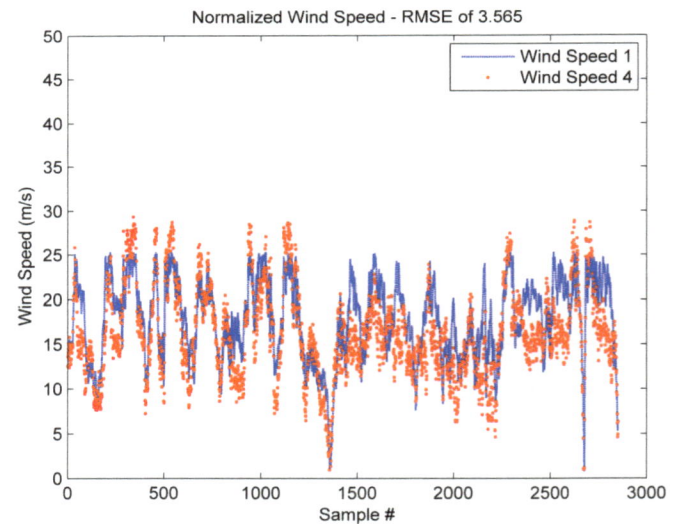

Figure 5. Shear Training Set 1 - Normalized Wind Speed Mean Signals

In order to quantify the differences in the wind speed measurements, the Root Mean Square Error (RMSE) is shown in each plot. The RMSE can be calculated between two anemometers by using Eq. (2), in which N is the number of samples in a data file and u_1 and u_2 are the wind speed mean values for the two anemometers considered in the calculation (Mohandes, Rehman, & Halawani, 1998).

The RMSE value for the normalized wind speed data in Figure 5 is much smaller than the RMSE value for the raw data in Figure 4; indicating that the normalization provided some measure of correcting for the different anemometer elevations.

$$RMSE = \left(\frac{1}{N} * \sum_{i=1}^{N} \left(u_1(i) - u_2(i) \right)^2 \right)^{1/2} \qquad (2)$$

5. RESIDUAL BASED FEATURE EXTRACTION

5.1. Difference Signal

The residual based feature extraction for the paired data set does not involve any data normalization nor does it use auto-associative neural network models. The difference signals are used as a surrogate for the residuals and are defined by Eq. (3) and Eq. (4). They are simply the difference between the wind speed mean signals (u_1 and u_2) for the paired anemometers. The k-means based figure of merit calculation further processes these two difference signals to determine the health state of both anemometers.

$$d_{12} = u_1 - u_2 \qquad (3)$$

$$d_{21} = u_2 - u_1 \qquad (4)$$

An example of anemometers in a nominal healthy condition is illustrated in Figure 6, in which the wind speed mean values for each sample are matching very closely. The further processing of this data file by the difference signal and the figure of merit calculation resulted in this data file being classified in the healthy condition.

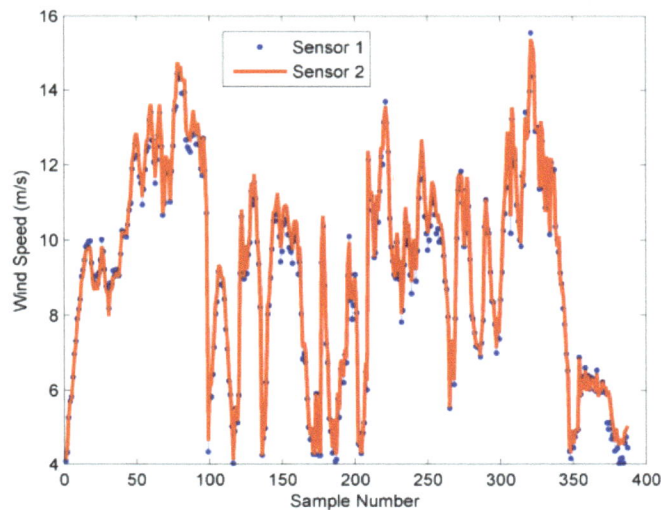

Figure 6. Wind Speed Mean Signals for Paired Test File 2 – Example of Anemometers in Nominal Healthy Condition

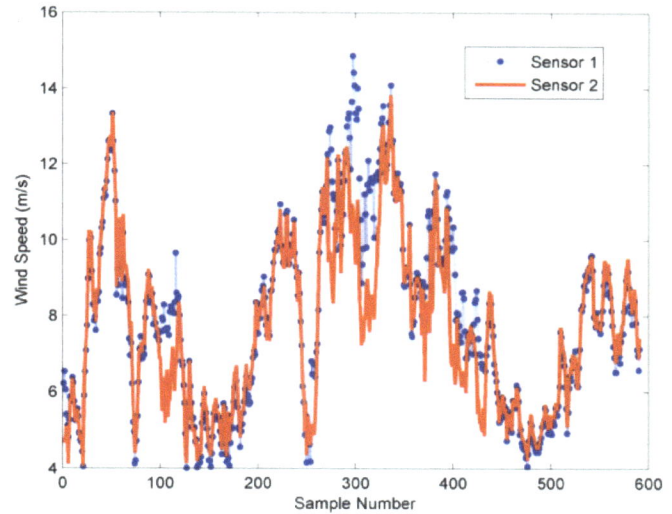

Figure 7. Wind Speed Mean Signals for Paired Test File 25 – Example of Second Anemometer Reading Slower

The signature that is exhibited when one of the paired anemometers is not working properly is highlighted in Figure 7. In this example, there are significant differences in the mean wind speed values for the two paired anemometers. However, these large differences are observed for only a portion of the samples. The observation that the signature only appears for a portion of the samples provides the motivation for clustering the difference signal and calculating a metric based on the cluster that contains information on the lagging sensor.

5.2. Auto-Associative Residual Processing

When a dynamic model of the system is not available a priori, the use of data driven health monitoring algorithms becomes a suitable alternative for monitoring the system health state (Schwabacher, 2005). Although there are various regression or distance from normal based metrics that are available, the use of auto-associative neural network (AANN) has some intriguing characteristics that make it particular suitable for this application. Its ability to learn non-linear correlation relationships and calculate residual values for each sensor provides a means to calculate a system health value. In addition, contribution plots for each sensor can also be used to provide diagnostic information (Thissen, Melssen, & Buydens, 2001). These attractive attributes of an auto-associative neural network have seen its usage for health monitoring span a diverse set of applications; from diesel engines (Antory, Kruger, Irwin, & McCullough, 2005) sensor health diagnostics and calibration (Xu, Hines, & Uhrig, 1999), to commercial aircraft engines (Hu, Qiu, & Iyer, 2007).

The theory and mathematics for the AANN were first described by Kramer (1991) and this method is effectively a way to perform non-linear principal component analysis.

Although principal component analysis (PCA) has been used in a variety of applications for process monitoring by using Hotellings' T^2 statistic and the residual square prediction error statistic (SPE); its assumption of the signals being linearly correlated is not satisfied in many engineering systems. An auto-associative neural network provides a similar framework, but has the ability to learn the non-linear correlation relationship among sensor variables. In this application, the underlying correlation relationship between the shear anemometer sensors is potentially non-linear; this is suggested by the power or logarithmic equations used to relate wind speed height and speed. The auto-associative neural network is applied after data normalization is done to correct for the wind speed height. However, the difference in the anemometer wind speed values in Figure 5 implies that the underlying relationship is not completely described by the power law. An auto-associative neural network can be used to further learn the sensor correlation relationship.

The AANN model structure consisted of 5 layers, an input layer, mapping layer, bottle neck layer, de-mapping layer, and an output layer as shown in Figure 8. One of its interesting aspects is that the network is trained with the same inputs and targets and thus the network is performing an identity mapping in which the output layer is providing an approximation of the inputs. The structure of the network used in this study follows the suggested configuration provided by Kramer (1991) and consists of 4 transfer functions. In sequential order, they consist of a tan-sigmoid transfer function, a linear transfer function, a tan-sigmoid transfer function, and a linear transfer function.

associative neural network models for the 3 or 4 sensor shear anemometer configurations. The inputs for the network consisted of the wind speed mean, maximum, and minimum values for each anemometer; this provided 12 and 9 inputs for the 3 and 4 shear anemometer configurations respectively. The structure of the AANN model used in this study was configured so that the numbers of nodes in the mapping layer were the same as the number of nodes in the de-mapping layer. The number of mapping and de-mapping nodes consisted of 5 and 7 for the 3 and 4 anemometer configurations. In both configurations, the bottle neck layer consisted of 3 nodes. The number of bottleneck nodes represents the intrinsic dimension of the data in a similar sense to the number of principal components retained in linear PCA (Kramer, 1991).

As an additional extension of using the AANN models for anemometer health assessment, it was postulated that it might be advantageous to have an ensemble of training models. This could provide a way of giving more weight to training models that provide a more accurate sensor prediction for a given anemometer shear test data file. The rationale for considering this aspect is that there are several un-modeled sources of variation. Variation due to manufacturing, site topography, and installation, could potentially impact the AANN model accuracy.

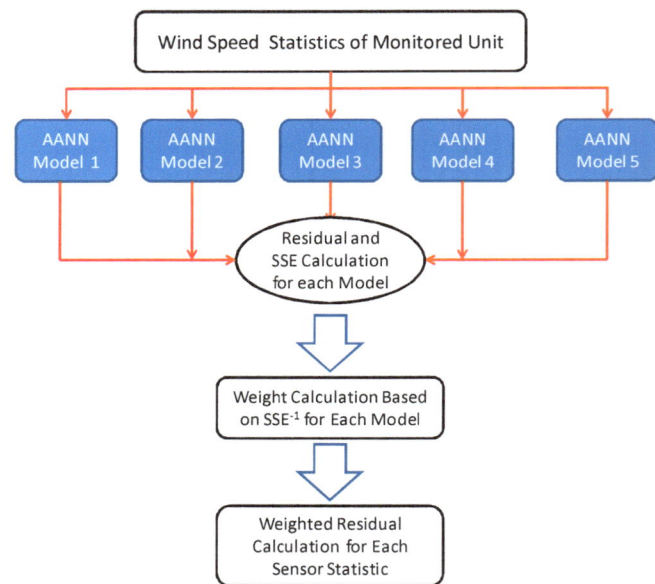

Figure 9. Flow Chart of Weighted Residual Calculation

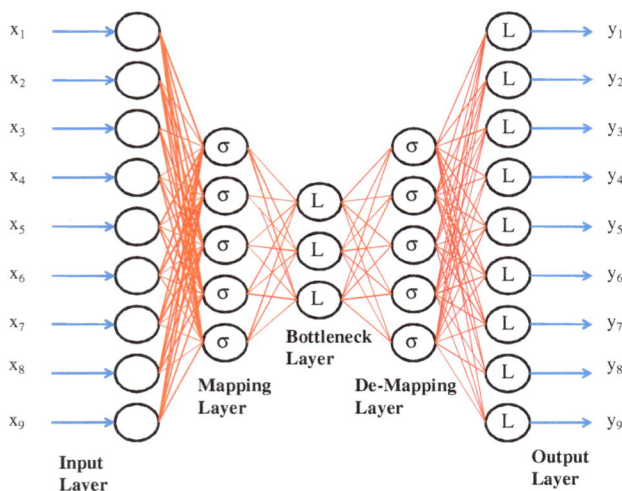

Figure 8. Auto-Associative Neural Network (9-5-3-5-9), σ is for tan-sigmoid transfer functions, L for linear transfer functions, x are inputs to the network and y are outputs of the network

Although the network structure uses the same transfer functions, there were some minor differences in the auto-

Using a weighted approach allows one to weight training models that might more closely represent the test data set. This can reduce variances due to other factors and allows one to assume that the deviation from the model is due to anomalous anemometer sensor behavior. A conceptual diagram of the weighted residual approach is highlighted in

Figure 9 and the details of the calculation procedure are further described in this section.

The baseline files provided for the shear anemometer data set consisted of two training baseline files for the three anemometer configuration and five training baseline files for the 4 anemometer configuration. The weighted AANN residual approach consisted of having 7 trained AANN models for each of the baseline files, with 2 being assigned to the three anemometer configuration and 5 assigned to the 4 anemometer configuration. For a given test file, the residuals for each anemometer sensor statistic would be calculated for each model that matched the anemometer configuration for a given test file.

The residuals for each sensor statistic are weighted by a weight vector that is calculated from the sum of square error value as shown in Eq. (5) and Eq. (6). In this calculation, SSE_k is the sum of square error for k^{th} AANN model, and r_{ijk} is the residual based on the predicted AANN value and the actual sensor statistic value for the ith data sample, the jth sensor, and the kth AANN model. In addition, N is the number of samples in the data file, and p and m is the number of input parameters and AANN models respectively. The weight for each model is calculated by taking the models SSE_k value and dividing that quantity by the summation of all the reciprocal SSE_k values. The weighted residual is then calculated by taking the weights for each model multiplied by the residuals as shown in Eq. (7). This provides a residual value for each sensor statistic that includes aspects from each training model, but provides more weight in training models that more closely match the test data set.

$$SSE_k = \sum_{j=1}^{p} \sum_{i=1}^{N} \left(r_{ijk}^{\,2} \right) \quad \text{for } k = 1 \text{ to } m \qquad (5)$$

$$w_k = \frac{(SSE_k)^{-1}}{\left(\sum_{k=1}^{m} SSE_k \right)^{-1}} \qquad (6)$$

$$wr_{ij} = \sum_{k=1}^{m} r_{ijk} w_k \quad \text{for } i = 1 \text{ to } N \text{ and } j = 1 \text{ to } p \qquad (7)$$

In order to evaluate the generalization of the AANN models and the weighted residual processing, the baseline data sets were randomly divided into a training set and a calibration set, in which the training set consisted of 70% of the available samples in a given baseline data file. An example of how well the predicted sensor statistic values match the actual values are shown in Figure 10 for the first shear baseline data file. In this example, the blue curve represents

the weighted predicted value from the AANN models and the red samples are the actual wind speed mean values. A measure of the model fit can be assessed by the root mean square error value (RMSE). In this example, the RMSE value is significantly lower when the AANN models are used as opposed to the results in Figure 5 that were obtained with only data pre-processing and normalization.

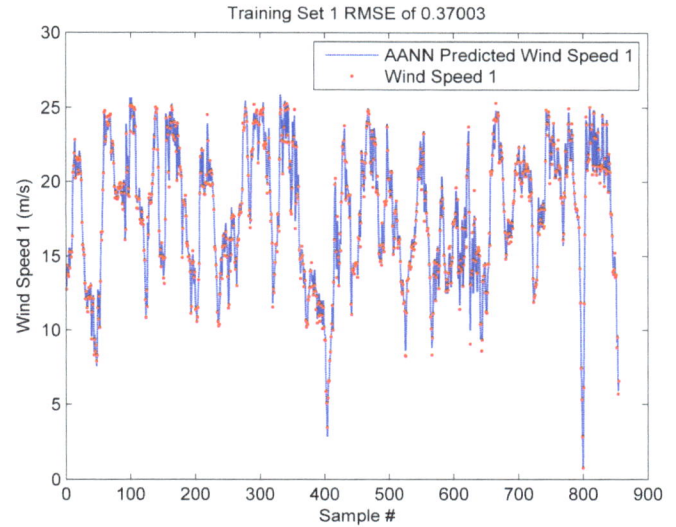

Figure 10. Shear Training Set - AANN Predicted and Measured Wind Speed Mean Values

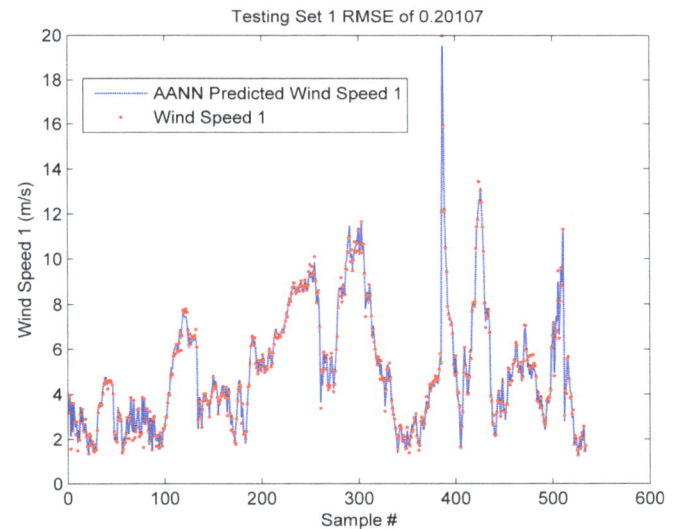

Figure 11. Wind Speed Predicted and Measured Value– Anemometers in Nominal Healthy Case (Shear Test File 1)

The trained AANN models and weighted residual processing method were then applied to the shear test files; example plots are shown in Figure 11 and Figure 12. In Figure 11, the results are for the first shear testing in which the predicted anemometer 1 wind speed mean and the actual anemometer 1 wind speed value are shown. Notice that the predicted and actual values match for the entire data set and the RMSE value is quite low. This is an example file in which the anemometers were considered to be in a healthy state.

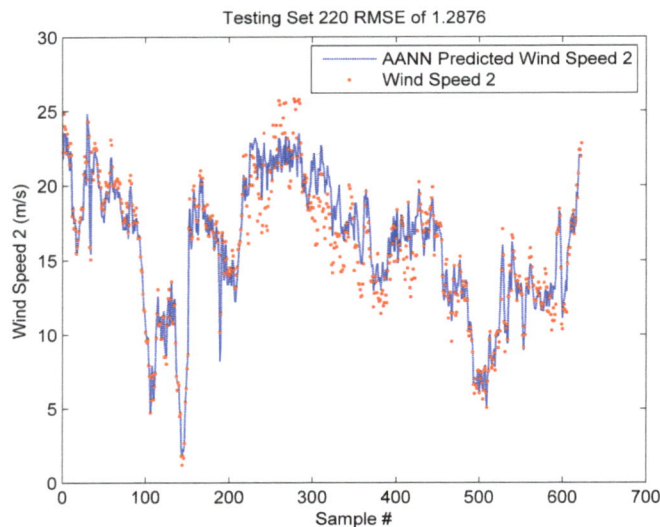

Figure 12. Wind Speed Predicted and Measured Value - Detected Faulty Case (Shear Test File 220)

An example of the anemometers in a degraded state is provided in Figure 12. In this example, there is a noticeable difference in the predicted and measured anemometer wind speed mean values for the second anemometer. The RMSE value for this case is also quite high. The bimodal fault signature is also observed, since the sensor is only lagging for a portion of the data samples. This highlights the motivation for clustering the residual signal, since the signature is only present for a particular subset of the operating conditions.

6. FIGURE OF MERIT

6.1. K-means Clustering

There are a variety of techniques used in data mining and artificial intelligence for clustering and density estimation (Jain, Murty, & Flynn, 1999). In this study, the k-means algorithm was used for partitioning the residual or difference wind speed mean values into two clusters. Density estimation using Gaussian mixture modeling was originally considered; however, the computation time became burdensome given the number of data files that had to be processed, and the k-means algorithm provided a more efficient way of determining the data clusters. The k-means clustering algorithm aims to partition the data set into a set of n clusters, where n is the number of clusters specified and its objective function is to minimize the within cluster sum of squares (Pollard, 1981). The algorithm is iterative in

nature, in that it is initialized with a random set of centroids and through the iteration process updates the center locations in order to reduce the within cluster sum of squares distance. The interested reader is referred to the work by Hartigan and Wong (1979) for a more detailed description of the k-means algorithm.

Although the k-means clustering does not guarantee a global solution, 5 replications are used in this study in order to select the lowest local minimum that is obtained for the 5 replications. The mean value is calculated in each cluster and the minimum value of the two clusters is stored and denoted as the figure of merit value. There is an additional logic rule to prevent a small sample cluster from being included. If the sample size of one of the two clusters is below 60 samples, the mean of the other cluster is stored as the figure of merit value. A small cluster could be due to a small amount of outlier samples that potentially made it through the data filtering screening. The motivation for selecting the cluster with the minimum mean value is based on the prior literature that suggest that a degraded anemometer would be reading slower than normal (Clark et al, 2009).

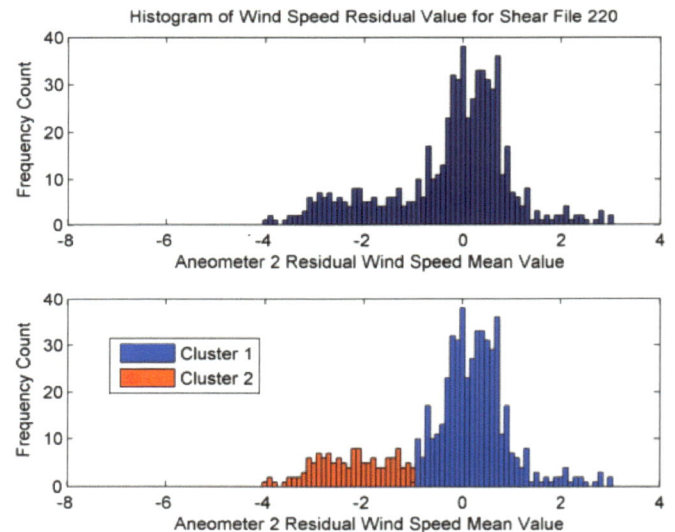

Figure 13. Wind Speed Residual Histogram and K-Means Clustering Result –File 220 Shear Data Set

An example of the k-means clustering result is provided in Figure 13. This result is from the residual wind speed signal for the system in a degraded health state. The histogram of the residual wind speed shows a bi-modal distribution in the top plot; the clustering result in the bottom graph indicates the two clusters that were determined using the k-means clustering. Considering that the figure of merit value is based on the mean value of the smaller of the two clusters, the k-means clustering provide a way of focusing on the samples when the anemometer is lagging. If one were to calculate statistics on the entire distribution without clustering, the algorithm would be less sensitive to the fault signature.

6.2. Figure of Merit Results

The previous section described how the figure of merit values were processed for both the shear and paired data sets; however the algorithm ultimately has to provide a decision statement on the health condition of each file. This required setting thresholds for the shear and paired figure of merit values. The literature suggests that an anemometer that is experiencing an increased level of friction and reading slower than normal could have an error of 1.5% to 3.0% and sometimes as high of a bias as 6% (Hale et al, 2011). The thresholds were based on selecting a value within that error range. The figure of merit thresholds for the shear anemometers were set at -0.35m/s for the three highest anemometers and a threshold of -0.5m/s for the anemometer at the lowest elevation. The anemometer at the lowest elevation was set with a more conservative threshold since it was believed that the AANN predicted values had more error for this anemometer. One should note that many of the shear files only had 3 anemometers, so in many instances only the first 3 thresholds are used. Considering that a fault is based on a lagging anemometer, a fault is declared if any of the figure of merit values are below its threshold and healthy otherwise. An example result for the figure of merit values is provided in Figure 14; this result is for the first anemometer for the shear data set. In this example, one can observe that the majority of the files are above the threshold. In total, 62 of the 255 shear anemometers were considered to be in a faulty state.

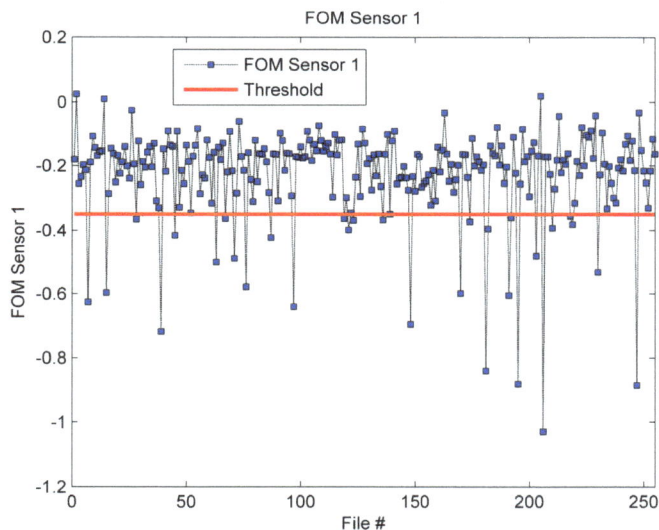

Figure 14. Figure of Merit Results for Shear Data Set

The figure of merit thresholds for both paired anemometers were set at -0.375m/s respectively. If the figure of merit value for a paired anemometer is below the threshold, that anemometer is considered in a failed condition and healthy otherwise. Although the algorithm is based on the difference signal and detecting degraded behavior for one of the anemometers; there were a few occurrences when the algorithm detected that both anemometers were in a failed state. This can occur if both anemometers are lagging but not in the same operating regime regarding wind speed or direction.

The figure of merit results for the paired data set is shown in Figure 15. The results show that the majority of files for the paired data set are detected in a healthy state. The paired health assessment algorithm detected 50 files with a degraded first anemometer, 43 files with a degraded second anemometer, and 325 files were classified as being in a healthy state. Only 2 files were detected as having both paired anemometers in a degraded state. This could be an indication that the algorithm was only suited for anemometer fault detection if there is at least one anemometer in a baseline state.

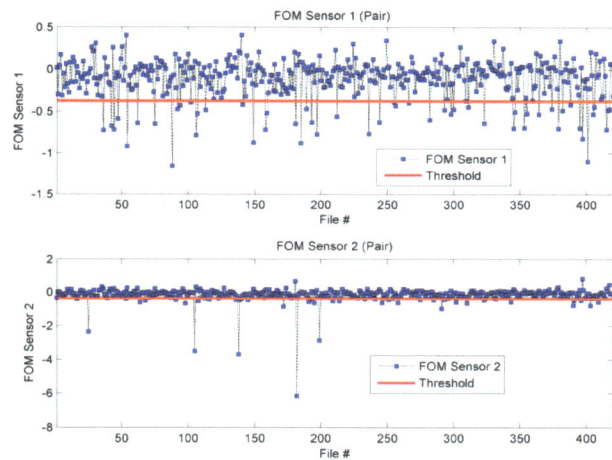

Figure 15. Figure of Merit Results for Paired Data Set

7. CONCLUSIONS

This paper introduced a health assessment methodology for assessing the condition of anemometers in two different configurations. The methodology consisted of a series of algorithmic processing steps from data filtering, to a residual calculation, to a k-means figure of merit health value. Although the algorithms for the shear and paired data sets were quite similar, the use of an auto-associative neural network and additional data normalization were performed by the shear health assessment algorithm. The algorithms for the paired and shear data sets resulted in the most accurate results for the Prognostics and Health Management Society 2011 Data Challenge. This highlights its potential merits for anemometer fault detection. In a general sense, this algorithm could be applied to many other sensor health monitoring applications. In particular, the use of auto-associative neural networks and the k-means clustering approach would be advantageous when redundant sensors are not available and the sensors can have an intermittent fault signature. The weighted residual processing using multiple baseline models is also useful for handling unit to unit variances since it weights training models that more accurately match the monitored unit. Extension of this health assessment algorithm can be approached in two directions; refinement for the specific case of anemometer fault detection, and also reconfiguring the algorithm for other applications.

8. SUGGESTIONS FOR FUTURE WORK

The proposed health assessment algorithm for the shear and paired anemometers provided encouraging results and there are several refinements considered for future work. The algorithm used for assessing the paired anemometers was tuned for the situation in which at least one of the two anemometers was in a healthy state. The inclusion of the wind speed variance information is being considered for future work in developing an algorithm that can detect that both paired anemometers are in a degraded state. This would provide a necessary extension to the proposed framework and would provide a way of detecting sensor problems without assuming that at least one of the reference measurements is in a healthy state.

Regarding refinements in the individual processing modules, a natural staring place would be the data filtering and normalization steps. These ultimately provide the inputs for all further processing, and improvement in removing samples due to icing or outlier values would likely aid the algorithms health monitoring accuracy. The use of an auto-associative neural network for calculating residuals can be compared with other residual processing methods, including the use of kernel principal component analysis methods as well as the traditional PCA methods. Also, the weighted residual processing method could be compared to selecting the top 1 or 2 models; the method for fusing the residual values is one area for further research. Although k-means was initially used for clustering the residual signal, density estimation using a mixture of Gaussians or other clustering techniques can be considered. In addition, evaluation of this proposed algorithm on other applications would further test its ability to generalize and work for other engineering systems.

ACKNOWLEDGEMENT

The work is supported by the NSF Industry/University Cooperative Research Center on Intelligent Maintenance systems (IMS) at the University of Cincinnati as well as its company members.

REFERENCES

Antory, D., Kruger, U., Irwin, G.W., & McCullough, G. (2005). Fault Diagnosis in Internal Combustion Engines using Nonlinear Multivariate Statistics. Proceedings of IME-Part I: Journal of Systems and Controls Engineering, vol. 219, pp. 243-258. doi: 10.1243/095965105X9614

Beltran, J., Llombart, A., & Guerrero, J.J. (2009). Detection of Nacelle Anemometers Faults in a Wind Farm. Proceedings of the International Conference on Renewable Energies and Power Quality, April 15-17, Valencia, Spain.

Capriglione, D., Liguroi, C., Pianese, C., & Pietrosanto, A. (2003). On-line Sensor Fault Detection, Isolation, and Accommodation in Automotive Engines. IEEE Transactions on Instrumentation and Measurement, vol. 52, pp. 1182-1189. doi: 10.1109/TIM.2003.815994

Clark, S.H., Clay, O., Goglia, J.A., Hoopes,T.R., Jacobs, L.T., Smith, R.P. (2009). Investigation of the NRG # 40 Anemometer Slowdown, Proceedings of the AWEA Windpower Conference and Exhibition, May 4-7, Chicago, IL.

Hale, E., Fusina, L., & Brower, M. (2011). Correction factors for NRG #40 Anemometers Potentially affected by Dry Friction Whip: Characterization, Analysis, and Validation. Wind Energy, doi: 10.1002/we.476

Hartigan, J.A., & Wong, M.A. (1979). Algorithm AS 136: A K-Means Clustering Algorithm, Journal of the Royal Statistical Society. Series C (Applied Statistics), vol. 28, pp. 100-108.

Hines, J.W., & Garvey, R.D. (2006). Development and Application of Fault Detectability Performance Metrics for Instrument Calibration Verification and Anomaly Detection. Journal of Pattern Recognition Research, vol. 1, pp. 2-15.

Hsu, S.A., Meindl, E.A., & Gilhousen, D.B. (1994). Determining the Power-Law Wind Profile Exponent under Near-Neutral Stability Conditions at Sea. Journal of Applied Meteorology, vol. 33, pp. 757-772.

Hu, X., Qiu, H., & Iyer, N. (2007). Multivariate change detection for time series data in aircraft engine fault diagnostics. IEEE Conference on Systems, Man, and Cybernetics, October 7-10, Montreal, Quebec. Doi: 10.1109/ICSMC.2007.4414131

Jain, A.K., Murty, M.N., & Flynn, P.J. (1999). Data Clustering: A Review. ACM Computing Surveys, vol. 31, pp. 264-323.

Kramer, M.A. (1991). Nonlinear Principal Component Analysis using Autoassociative Neural Networks. AIChe Journal, vol. 37, pp. 233-243. doi: 10.1002/aic.690370209

Kusiak, A., Zheng, H., & Zhang, Z. (2011). Virtual Wind Speed Sensor for Wind Turbines. Journal of Energy Engineering, vol. 137, pp. 59-69. doi: 10.1061/(ASCE)EY.1943-7897.0000035

Mohandes, M.A., Rehman, S., & Halawani, T.O. (1998). A Neural Networks Approach for Wind Speed Prediction. Renewable Energy, vol. 13, pp. 345-354. doi: 10.1016/S0960-1481(98)00001-9

Murakami, S., Mochida, A., & Kato, S. (2003). Development of Local Area Wind Prediction System for Selecting Suitable Site for Windmill. Journal of Wind Energy and Industrial Aerodynamics, vol. 22, pp. 679-688. doi: 10.1016/j.jweia.2003.09.040

Patton, R.J. (1991). Fault detection and diagnosis in aerospace systems using analytical redundancy, Computers & Control Engineering Journal, vol. 2, pp. 127-136.

Petersen, E.L., Mortensen, N.G., Landberg, L., Hujstrup, J., & Frank, H.P. (1998). Wind Power Meteorology Part II: Siting and Models. Wind Energy, vol. 1, pp. 55-72.

Peterson, E.W., & Hennessey Jr., J.P. (1977). On the Use of Power Laws for Estimates of Wind Power Potential. Journal of Applied Meteorology, vol. 17, pp. 390-394.

PHM Society 2011 Data challenge Competition, (2011). [http://www.phmsociety.org/competition/phm/11]

Pollard, D. (1981). Strong Consistency of K-Means Clustering. The Annals of Statistics, vol. 9, pp. 135-140. doi: 10.1214/aos/1176345339

Schwabacher, M. (2005). A Survey of Data Driven Prognostics. In AIAA Infotech@ Aerospace Conference, September 26-29, Arlington, VA.

Thissen, U., Melssen, W.J., & Buydens, L.M.C. (2001). Nonlinear Process Monitoring using bottle-neck neural networks. Analytica Chimica Acta, vol. 446, pp. 371-383. doi:10.1016/S0003-2670(01)01266-1

Venkatasubramanian, V., Rengaswamy, R., Yin, K., & Kavuri, S.N. (2003). A Review of Process Fault Detection and Diagnosis Part I: Quantitative Model-Based Methods. Computers and Chemical Engineering, vol. 27, pp. 293-311. doi: 10.1016/S0098-1354(02)00160-6

Xu, X., Hines, J.W., & Uhrig, R.E. (1999). Sensor Validation and Fault Detection using Neural Networks. In Proceedings of the Maintenance and Reliability Conference, May 10-12, Gatlinburg, TN.

David Siegel is currently a Ph.D. student in Mechanical Engineering at the University of Cincinnati and a research assistant for the Center for Intelligent Maintenance Systems. Related work experience in the field of prognostics and health management include internships at General Electric Aviation and at the U.S Army Research Lab. His current research focus is on component-level prognostic methods as well as health monitoring algorithms for systems operating under multiple operating regimes.

Jay Lee is an Ohio Eminent Scholar and L.W. Scott Alter Chair Professor at the University of Cincinnati and is founding director of National Science Foundation (NSF) Industry/University Cooperative Research Center (I/UCRC) on Intelligent Maintenance Systems which is a multi-campus NSF Center of Excellence between the University of Cincinnati (lead institution), the University of Michigan, and Missouri University of Science and Technology. His current research focuses on autonomic computing, embedded IT and smart prognostics technologies for industrial and healthcare systems, design of self-maintenance and self-healing systems, and dominant design tools for product and service innovation. He is also a Fellow of ASME, SME, as well as International Society of Engineering Asset Management (ISEAM).

A Multiple Model Prediction Algorithm for CNC Machine Wear PHM

Huimin Chen [1]

[1] *Department of Electrical Engineering, University of New Orleans, New Orleans, LA, 70148, U. S. A.*
hchen2@uno.edu

ABSTRACT

We present a multiple model approach for wear depth estimation of milling machine cutters using dynamometer, accelerometer, and acoustic emission data. The feature selection, initial wear estimation and multiple model fusion components of the proposed algorithm are explained in details and compared with several alternative methods using the training data. The performance evaluation procedure and the resulting scores from the submitted predictions are also discussed.

1. Introduction

The 2010 PHM data challenge focuses on the remaining useful life (RUL) estimation for cutters of a high speed CNC milling machine using measurements from dynamometer, accelerometer, and acoustic emission sensors (See http://www.phmsociety.org/competition/phm/10). The challenge data set contains six individual cutter records, denoted by c1, ..., c6. Records c1, c4 and c6 are training data while records c2, c3 and c5 are testing data. Each cutter was used repeatedly to cut certain work piece with the spindle speed of 10400 RPM. The wears of three flutes were measured after each cut (in 10^{-3}mm). In addition, 3-component platform dynamometer was used to measure the cutting forces in X, Y, Z directions. Three accelerometers were used to measure the vibrations of the cutting process in X, Y, Z directions. Acoustic emission (AE) sensor was used to measure the acoustic signature (AE-RMS) of the work piece during the cutting process. Prognostic algorithm development with similar equipment setup was reported in (Li et al., 2009). The training data contain 315 cut files for each cutter with the measured time series of forces, vibrations and AE-RMS and the resulting wears of three flutes after each cut. The testing data only contain the force, vibration and AE-RMS measurements for each cut without the wear depth measurement of each flute. The goal is to estimate the maximum number of cuts one can safely make for each testing cutter at a given wear limit. Note that one has to implicitly or explicitly predict the maximum wear of the flutes after each cut without knowing the initial wear of the cutter using only the force, vibration and AE-RMS measurements from consecutive cuts. Upon completion of the competition, the author was invited to submit a paper that fully discloses the algorithm used in 2010 PHM data challenge.

The rest of the paper is organized as follows. In Section 2, we first explain the performance evaluation criterion of the data challenge. Then we discuss feature selection for linear regression model on the additional wear after each cut for all three cutters. Finally, we reveal the need of individual regression model for each cutter and call for a multiple model approach to predict the wear depth of the testing cutter. In Section 3, we present the detailed description of the algorithm applied to the data challenge. The concluding summary is in Section 4.

2. Preliminary Analysis of the Training Data

2.1 Data Challenge Submission and Score Function

Each participant in 2010 PHM data challenge is required to estimate the maximum safe cuts at wear limit of 66, 67, ..., 165 ($\times 10^{-3}$mm) for three testing cutters. However, one does not know the actual wear after each cut and can only use the sensor measurements up to the current cut to infer the cumulative wear (although a participant can use the measurements from all 315 cuts, which is clearly unrealistic in practice). Note that the wear limit is on the maximum wear among three flutes. From the training data, we obtained the maximum safe cuts for the three training cutters in Figure 1. Clearly, c1 and c4 have early jump in the number of maximum cuts before wear depth reaches 100 while c6 has a small jump before wear depth reaches 120. These are important change points where the wear depth of the cutter is small after each cut (and in some cases unnoticeable) so that the number of safe cuts increases abruptly when the wear limit changes incrementally. Note that the initial wears of c1, c4, and c6 are 48.9, 31.4,

62.8, respectively.

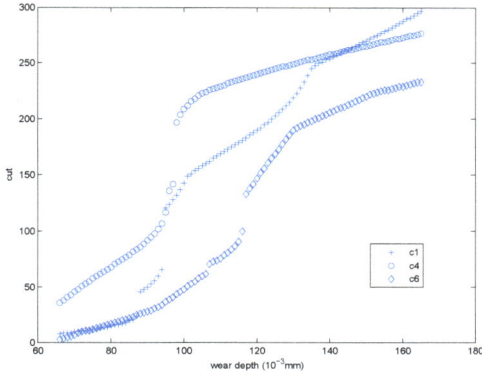

Figure 1. Maximum safe cuts for c1, c4 and c6.

Denote by δ the difference between the estimated number of cuts and the actual number of maximum cuts for a given cutter before it reaches the wear limit. The score function is defined as

$$S(\delta) = \begin{cases} e^{-\delta/10} - 1 & \delta < 0 \\ e^{\delta/4.5} - 1 & \delta \geq 0 \end{cases} \quad (1)$$

The goal is to minimize the sum of scores at the given 100 wear limits for all three testing cutters. Note that over estimate of the cut number has a larger penalty than under estimate. The above exponential penalty function focuses more on penalizing any single bad estimate rather than the average performance among all possible wear limits.

2.2 Feature Extraction and Selection

We first consider a regression model to estimate the additional wear after each cut assuming that the initial wear of the cutter is known. The goal is to identify useful features and to avoid overfitting to the training data. Linear regression was used with the candidate features and their p-values resulting from the regression on c1, c4, and c6 cutters in Table 2.2. A small p-value indicates the corresponding feature is likely to have non-zero regression coefficient, i.e., it should be included in wear depth estimation (Schervish, 1996). Unfortunately, most of the candidate features have relatively small p-values, making feature selection a challenging task with limited training data. Note that one can explore many candidate features including peak, mean, standard deviation, skew, kurtosis of force and vibration along X, Y, Z directions as well as their frequency domain indicators as listed in Table 2.2. Thus it is computationally infeasible to exhaustively enumerate all possible feature subsets with the linear regression model and choose the best one. To expedite the automatic feature selection procedure for any regression model, we do not want to test whether an individual regression coefficient should be zero or not based on p-value, instead, we control the false discovery rate (FDR) among all the selected features. It is an effective method for testing multiple hypotheses simultaneously (Benjamini & Hochberg, 1995).

Maximum force	0.08
Mean force	0.12
Standard deviation of force	0.07
Maximum vibration	0.05
Mean vibration	0.06
Standard deviation of vibration	0.09
Maximum AE-RMS	0.06
Mean AE-RMS	0.08
Standard deviation of AE-RMS	0.15
Peak magnitude of force frequency spectrum below 2000Hz	0.11
Peak magnitude of vibration frequency spectrum below 2000Hz	0.07
Peak magnitude of AE-RMS frequency spectrum above 2000Hz	0.14

Table 1. Candidate features and the corresponding p-values in linear regression

Formally, we consider d candidate features to be possibly included in the linear regressor. A hypothesis H_k describes the index set $\mathcal{I}_k \subseteq \{1, \cdots, d\}$ of the non-zero regression coefficients of $\mathbf{w}$, i.e., the selected feature subset.

H_k: $w_i \neq 0$ if $i \in \mathcal{I}_k$, otherwise $w_i = 0$, $i = 1, \cdots, d$.

We apply the FDR control technique to estimate $\mathcal{I}_k$. Note that the procedure does not require any independence assumption of the test statistics which is important in our case since some candidate features can have strong correlations in the regression model. The feature selection procedure is a step-down test (by successively selecting features) which is more efficient than a step-up test (by successively eliminating non-diagnostic features) when the number of selected features is relatively small compared with d. The procedure starts with the test statistic $T_1, \cdots, T_d$ based on the element-wise estimate $\hat{w}_1, \cdots, \hat{w}_d$. Each test statistic T_i is associated with a p-value, π_i, indicating its statistical significance when $w_i = 0$.

For any user specified FDR level $q \in (0, 1)$, the feature subset is selected by performing the following steps which controls the FDR to be below q (Chen, Bart, & Huang, 2008).

- Order the p-values such that $\pi_{(1)} \leq \cdots \leq \pi_{(d)}$.

- Compute the index $u_i = \min\left(1, \frac{d}{(d-i+1)^2}q\right)$, $i = 1, ..., d$.

- Reject all hypotheses $w_{(j)} = 0$ for $1 \leq j \leq k-1$ where k is the smallest index for which $\pi_{(k)} > u_k$. If no such k exists, then $\mathbf{w} = 0$.

Once the subset $\hat{\mathcal{I}}_k$ is determined, the regression coefficients should be recomputed using only the selected input features. Clearly, the FDR controlled feature selection procedure is much more efficient than finding the optimal feature subset via enumeration. It has been shown that the above procedure does control the FDR at the significance level q (Chen et al., 2008).

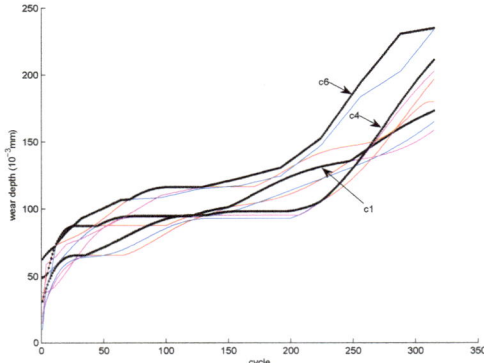

Figure 2. Flute wear after each cut and the maximum wear for cutter c1, c4, c6.

We set $q=0.1$ and applied linear regression to 68 candidate features. Only 5 features were selected based on the training data. With the selected features, we obtained the regressed additional wear for c1, c4 and c6 and the score being $3 \cdot 10^6$. We also tried to build the linear regression model for the wear of each flute instead of using the maximum wear after each cut. Figure 2 shows the individual flute wear and the maximum wear after each cut for the three training cutters. With the same FDR control level, the resulting score becomes $4 \cdot 10^6$ owing to the regression model fitting to the individual wear of each flute instead of the maximum wear. However, when we applied linear regression to c1 and c4 and the resulting model to estimate the wear of c6, the score becomes $2 \cdot 10^8$. Similar effects were observed when using c1 and c6 to build the model and estimate the wear of c4 as well as using c4 and c6 to estimate the wear of c1. Note that the results are based on the true initial wear of the cutter. It is clear that the linear regression method on the selected features tends to overfit even with fairly restrictive FDR control. The resulting score on the training set is unsatisfactory.[1] From the preliminary data analysis, we concluded that

- A single linear regression model yields large estimation error of the wear depth even with controlled FDR in feature selection.

- Regression on the additional wear after each cut for individual flute does not gain any benefit compared with regression on the maximum wear.

---

[1] At the time that I registered for 2010 PHM data challenge, the top score on the leader board was already $4 \cdot 10^5$ on testing data without knowing the initial wear on each cutter.

- No small subset of the candidate features can correlate well with the maximum wear after each cut for all training cutters.

3. MULTIPLE MODEL PREDICTION ALGORITHM

3.1 Building Regression Model for Each Cutter

We assume that each cutter has its own model to estimate the additional wear after each cut based on the selected features. In general, we consider a class of models $M_1, ..., M_K$ where model M_i assumes that the observation z is governed by a likelihood functional $f_i(z|\theta_i)$ depending on the unknown parameter θ_i ($i = 1, ..., K$). The dimension of θ_i is denoted by p_i. Denote by z^n a vector of n independent observations. Given z^n, one wants to find the best model M_i among the K candidates. Existing model selection criteria can be written in a general form (Chen & Huang, 2005)

$$l_j = -\log f_j(z^n|\hat{\theta}_j) + d_j(z^n), \; j=1, ..., K \qquad (2)$$

being minimized among the K candidates. The first term of l_j uses the best estimate of θ_j to fit the negative log-likelihood function. The second term $d_j(z^n)$ is a penalty function that varies for different model selection criteria.

From our past experience, we applied the minimum description length (MDL) criterion which yields $d_j(z^n)$ $=\log\left(\int f_j(z^n|\hat{\theta}_j)dz^n\right)$. It is interpreted as part of the normalized maximum likelihood (NML) (Rissanen, 1996). Under certain regularity conditions, one can use the asymptotic expansion of $d_j(z^n)$ given by

$$d_j(z^n) = \frac{p_j}{2}\log\left(\frac{n}{2\pi}\right) + \log\int |I(\theta_j)|^{1/2}d\theta_j \qquad (3)$$

where $I(\theta_j)$ is the Fisher information matrix given by

$$I(\theta_j) = \lim_{n\to\infty}\frac{1}{n}E\left[-\frac{\partial^2 \log f_j(z^n|\theta_j)}{\partial\theta_j(\partial\theta_j)'}\right] \qquad (4)$$

and the integral in (3) is over an appropriate subset of the parameter space. The MDL criterion intends to minimize the overall code length of a model and the observation described by the model.

As a special case of the above MDL principle, we assume that the additional wear of each cutter is a polynomial function of the selected features of unknown order. Thus for model M_i, the observation equation is $\mathbf{y} = \mathbf{H}_i\theta_i + \mathbf{w}_i$ where $\mathbf{H}_i$ is a known $n \times p_i$ matrix; θ_i is an unknown $p_i \times 1$ vector; and $\mathbf{w} \sim \mathcal{N}(0, \sigma^2\mathbf{I})$ is the Gaussian noise vector with unknown variance σ^2. The minimum variance unbiased (MVU) estimate of θ_i is $\hat{\theta}_i = (\mathbf{H}_i^T\mathbf{H}_i)^{-1}\mathbf{H}_i^T\mathbf{y}$. The residual sum of squares (RSS) of $\hat{\theta}_i$ is $R_i = ||\mathbf{y} - \mathbf{H}_i\hat{\theta}_i||^2$. The MVU estimate of σ^2 is $\hat{\sigma}_i^2 = R_i/(n - p_i)$, which is different from the ML estimate given by R_i/n. The MDL based on the NML density minimizes the cost (Rissanen, 1996)

$$\text{MDL}(p_i) = \frac{n}{2}\log\left(\hat{\sigma}_i^2\right) + \frac{p_i+1}{2}\log F_i + L_i \qquad (5)$$

where $F_i = (\mathbf{y}^T\mathbf{y} - R_i)/(p_i\hat{\sigma}_i^2)$ and $L_i = \frac{1}{2}\log\left(\frac{n-p_i}{p_i^3}\right)$. We control the FDR level $q=0.1$ as before. For cutter c1, 7 features were selected resulting in a score of $3 \cdot 10^3$ from the penalized regression model. Interestingly, 6 features were selected (4 of which are identical to those used for c1) for cutter c4 leading to a score of $2 \cdot 10^3$. 9 features were selected for cutter c6 (including all 6 featured used for c4) leading to a score of $2 \cdot 10^3$. We can see that three training cutters with different regression models have much better accuracy in estimating the wear depth than using a single regression model. However, we can no longer perform cross validation on the resulting models with the training data.

3.2 Multiple Model Fusion

From the Bayesian point of view, selecting a single model to make prediction or interpolation ignores the uncertainty left by finite data as to which model is the correct one. Thus a Bayesian formalism uses all possible models in the model space under consideration when making predictions, with each model weighted by its "posterior" probability being the correct one. The approach is called Bayesian model averaging and widely used in combining different learning algorithms (MacKay, 1992). The major difficulty in applying the Bayesian approach lies in the specification of priors when the candidate models are nested or partially overlapped. To be more specific, assuming that we have the observation vector z^n, the likelihood of a new observation y can be approximated by

$$f(y|z^n) \approx \sum_{i=1}^{K} P(M_i|z^n)f_i(y|\hat{\theta}_i) \qquad (6)$$

using multiple models while the single model likelihood is $f_i(y|\hat{\theta}_i)$ if model M_i is selected. Without assigning prior to θ_i, one can not apply Bayes formula to estimate the model probability.

If one does not stick to the formal Bayesian solution, then the penalty l_i in the model selection criterion can be used to estimate the model probability. We apply the following estimator

$$P(M_i|z^n) = e^{-l_i} / \left(\sum_{j=1}^{K} e^{-l_j}\right) \qquad (7)$$

and use MDL for l_i. In our opinion, the model probability can not be interpreted as the posterior since the prior of θ_i is unspecified. It represents the self-assessment of how likely M_i is selected using the criterion l_i. Denote by $\hat{y}_i$ the estimate (prediction or interpolation) of y using M_i. The single model estimate uses $\hat{y}_i$ if M_i is selected as the best model. The multiple model estimate uses

$$\hat{y} = \sum_{i=1}^{K} P(M_i|z^n)\hat{y}_i \qquad (8)$$

and a subset of the K models can be identified based on a predetermined minimum model probability. Note that the

outcome provided via multiple model fusion is valid for any model set, not limited to linear models. The essence of (7) is to approximate the Bayesian evidence of each model without any dependence on the unknown parameter θ_i so that all data can be used for the inference purpose, i.e., parameter estimation (Chen & Huang, 2005). If we know the initial wear of the testing cutter, then we can apply the above prediction method with three regression models obtained from the training data.

3.3 Initial Wear Estimate and RUL Prediction

Typically, the initial wear of the testing cutter should be known to the algorithm developer. Then for each cut, one collects the sensor measurements and sequentially predicts the additional wear after the cut. However, this information was unavailable during the competition. To estimate the initial wear, we made a hypothesis that the standard deviation of the high frequency peak from AE-RMS measurements correlates with the cutter's initial wear. This also has a statistical support with relatively small p-value from the linear regression model. We extrapolate from the three training cutters and obtained the estimated initial wears for c2, c3, c5 being 60, 55, 45, respectively. Note that we have used the AE-RMS measurements from the testing cutter of the first 15 cuts, which seems reasonable in practice.

With the estimated initial wear and the three regression models learned from the training data, we apply the multiple model algorithm via sequentially estimating the model probabilities after each cut by processing new sensor measurements and combining the individual predictions using (8). The resulting prediction of the maximum safe cuts is shown in Figure 3 where one can see that the prediction looks like a weighted average of the wears made by c1, c4, and c6. The first submission (with alias UNO-PHM) to 2010 PHM data challenge had a score of $9 \cdot 10^5$, which was among those top performance teams on the leader board.

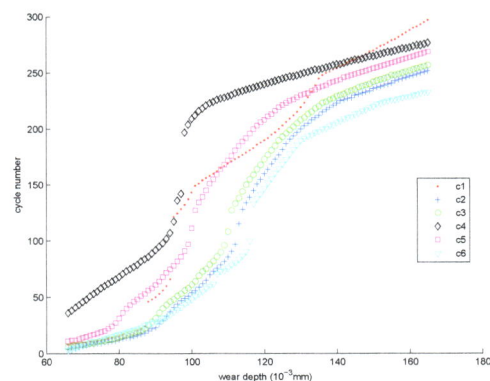

Figure 3. First submission to PHM data challenge for cutter c2, c3, c5 (maximum safe cuts of c1, c4, c6 included for comparison).

4. Conclusion

We provided detailed description of the multiple model prediction algorithm with automatic feature selection and initial wear estimation submitted to 2010 PHM data challenge. The final submission ranked #2 among professional and student participants and the method is applicable to other data driven PHM problems.

Acknowledgments

The author wants to express his gratitude to Dr. Neil Eklund for organizing the 2010 PHM data challenge and inviting him to the special session, to Dr. Kai Goebel for hosting his visit to NASA Ames Research Center, arranging stimulating discussions with his team members on various prognostic problems, and to the ijPHM editor, Dr. Abhinav Saxena, as well as anonymous reviewers for providing constructive comments on an earlier version of this paper.

He is also grateful to the research sponsorship in part by ARO through grant W911NF-08-1-0409, ONR-DEPSCoR through grant N00014-09-1-1169, and Louisiana Board of Regents through NASA EPSCoR DART-2 and NSF(2010)-LINK-50.

Nomenclature

δ	estimation error of maximum safe cuts
d	number of candidate features
w_i	regression coefficient of feature i
π_i	p-value of feature i
y	observation or quantity to be estimated
$\hat{y}$	estimate of y
M_i	statistical model i
θ_i	unknown parameter associated with model i
f_i	likelihood function of model i
l_i	penalized log-likelihood function of model i
z^n	n independent observations

References

Benjamini, Y., & Hochberg, Y. (1995). Controlling the False Discovery Rate - A Practical and Powerful Approach to Multiple Testing. *Journal of the Royal Statistical Society, B57(1)*, 289-300.

Chen, H., Bart, H., & Huang, S. (2008). Integrated Feature Selection and Clustering for Taxonomic Problems within Fish Species Complexes. *Journal of Multimedia, 3(3)*, 10-17.

Chen, H., & Huang, S. (2005). A Comparative Study on Model Selection and Multiple Model Fusion. In *International Conference on Information Fusion.*

Li, X., Lim, B., Zhou, J., Huang, S., Phua, S., Shaw, K., et al. (2009). Fuzzy Neural Network Modelling for Tool Wear Estimation in Dry Milling Operation. In *Annual Conference of the Prognostics and Health Management Society.*

MacKay, D. (1992). Bayesian Interpolation. *Neural Computation, 4*, 415-447.

Rissanen, J. (1996). Fisher Information and Stochastic Complexity. *IEEE Trans. on Information Theory, 42*, 40-47.

Schervish, M. (1996). P Values: What They Are and What They Are Not. *The American Statistician, 50(3)*, 203-206.

Huimin Chen received the B.E. and M.E. degrees from Department of Automation, Tsinghua University, Beijing, China, in 1996 and 1998, respectively, and the Ph.D. degree from the Department of Electrical and Computer Engineering, University of Connecticut, Storrs, in 2002, all in electrical engineering. He was a post doctorate research associate at Physics and Astronomy Department, University of California, Los Angeles, and a visiting researcher with the Department of Electrical and Computer Engineering, Carnegie Mellon University from July 2002 where his research focus was on weak signal detection for single electron spin microscopy. He joined the Department of Electrical Engineering, University of New Orleans in Jan. 2003 and is currently an Associate Professor. His research interests are in general areas of signal processing, estimation theory, and information theory with applications to target detection and target tracking.

Detection/Diagnosis of Chipped Tooth in Gears by the Novel Residual Technology

L. Gelman[1], I. Jennions[2], and I. Petrunin[3]

[1,3] *School of Engineering, Cranfield University, Cranfield, Bedfordshire, MK43 0AL, UK*
l.gelman@cranfield.ac.uk
i.petrunin@cranfield.ac.uk

[2] *IVHM Centre, Cranfield University, Cranfield, Bedfordshire, MK43 0AL, UK*
i.jennions@cranfield.ac.uk

ABSTRACT

The novel residual technology is applied for the detection/diagnosis of partly-missing (chipped) tooth in a gear of the machine fault simulator (MFS) produced by SpectraQuest (USA). The automated sensor-less technique is implemented for the speed estimation. This technique estimates the speed data from raw vibration data using the narrow-band demodulation of the mesh component, providing that an approximate running speed and number of teeth are known. An advanced technique based on the likelihood ratio is used for decision making. The novel technology is compared with the conventional technique, the classical residual technology. For both technologies, the gear fault has been continuously diagnosed throughout the whole test duration without false alarms and missed detections. The use of the novel residual technology in comparison to the classical residual technology provides higher probability of the correct damage detection and faster damage diagnosis.

1. INTRODUCTION

Three main approaches are applied for detecting faults in geared systems: acoustic analysis, debris monitoring and vibration analysis. The vibration-based diagnosis has been the most popular monitoring technique because of its high effectiveness. Local fault detection in gears using vibration analysis has been a subject of intensive research. A plethora of methods have been proposed including amplitude and phase demodulation (McFadden, 1986, Gelman et al., 2005, Combet, Gelman, Anuzis, & Slater, 2009), cepstrum analysis (Randall 1982), residual analysis (Combet at al., 2009, Wang, Ismail, & Golnaraghi, 2001), adaptive filtering (Brie, Tomczak, Oehlmann, & Richard, 1997, Lee & White, 1998, Combet & Gelman, 2009), use of a model

(Wang & Wong, 2002, Martin, Jaussaud, & Combet, 2004), inverse filtering (Lee & Nandi, 2000, Endo & Randall, 2007), time frequency (TF) analysis (Wang & McFadden, 1993, Forrester, 1996, Choy, Polyshchuk, Zakrajsek, Handschuh, & Townsend, 1996, Wang & McFadden, 1996, Loutridis, 2006, Halima, Shoukat Choudhuryb, Shaha, & Zuoc, 2008) and time-scale analysis (Wang et al., 2001, Dalpiaz, Rivola, & Rubini, 2000, Lin & Zuo, 2003). The majority of the gear fault detection methods are based on the residual signal as classically obtained after the removal of the mesh harmonic components from the gear vibration signal which is processed using the time synchronous average (TSA) (Stewart, 1977, McFadden, 1987). However, our literature search (McFadden, 1986, Combet at al., 2009, Wang at al., 2001, Lee & White, 1998, Halima et al., 2008, Stewart, 1977, McFadden, 1987) showed that nobody has investigated the effectiveness of the residual technology for detection of partly missing (chipped) gear tooth. Therefore, the aim of this paper is to investigate for the first time detection of partly missing (chipped) gear tooth by the residual technology.

2. THE NOVEL RESIDUAL TECHNOLOGY

Schematic of the damage detection technology, based on the novel residual signal is shown in Fig. 1. The first stage of the technology is the angular re-sampling of the raw vibration signal using the estimate of the shaft speed. The speed estimate can be obtained using the one / rev signal from a tachometer; however, in some cases, the speed can be extracted from vibration data without need of the tachometer signal. In the present paper, the advanced automatic technology for the time synchronous averaging of the raw gear vibrations has been employed (Combet & Gelman, 2007). This technology does not require detailed speed data; only the approximate value of the running speed and number of teeth are necessary.

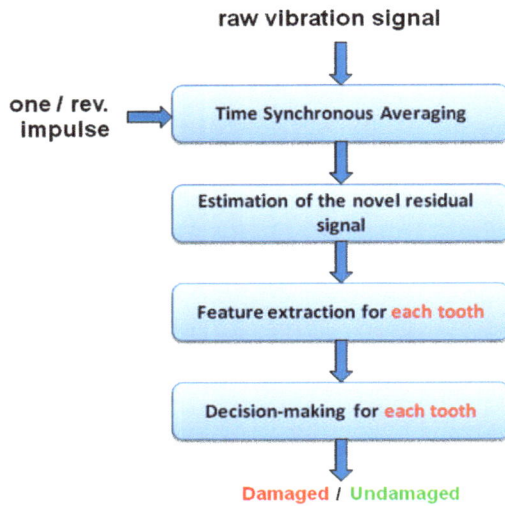

Figure 1. Schematic of the novel residual damage detection technology.

The speed is estimated from the automatically selected mesh component of the vibration data using the narrow-band demodulation (Combet & Gelman, 2007). The TSA is then performed for the signal re-sampled according to the estimated speed. In order to estimate the signal duration required for the TSA, the dependency of the ratio between the averaged variance of the TSA signal and the variance of the re-sampled vibration signal versus the number of averages was estimated. The length of the raw signal required for the TSA was estimated according to the number of averages at which the ratio of variances begin demonstrate the stationary-like behavior (Fig. 2). This approach is proposed in (Combet & Gelman, 2009).

The novel residual signal is obtained from the TSA signal by removing not only the mesh harmonics but also low and high shaft orders. This approach is proposed in (Combet at al., 2009). It improves detection effectiveness as impacts created by local gear faults usually affect the relatively narrow frequency range of the gearbox vibration spectrum.

The extraction of the diagnostic features is based on the averaging of the residual signal envelope within the meshing interval. The feature values are obtained for each tooth of the gear.

The decision making procedure is based on the estimation of the likelihood ratio using training data with *a priori* known classification. During the testing, values of the likelihood ratio are accumulated and compared to the threshold in order to make the final decision. This procedure allows for making decision for each tooth of the gear separately.

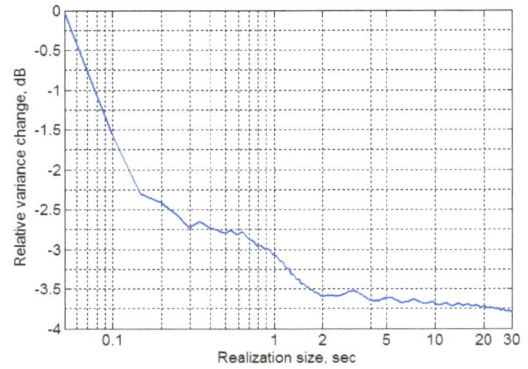

Figure 2. Estimation of the relative variance change of the TSA signal vs realization size.

3. THE EXPERIMENT DESCRIPTION

The experimental set up is based on the MFS test rig produced by SpectraQuest (USA) (Fig. 3) equipped with one-stage gearbox (Fig. 4). Two gears were used for the test: gear 1 with no damage on teeth and gear 2 with a partly missing (chipped) tooth (Fig. 5). The test was performed at a constant shaft speed of 3000 rpm (measured on the motor shaft) which corresponds to approximately 1200 rpm on the tested gear shaft.

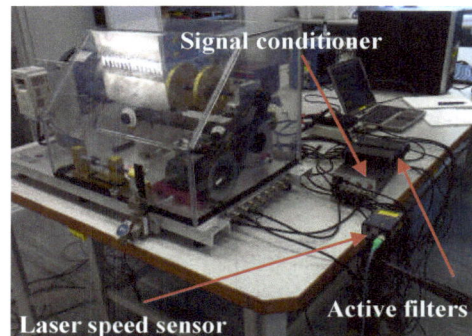

Figure 3. The test rig (MFS).

Figure 4. The gearbox with installed accelerometers.

Figure 5. The gear with partly missing tooth.

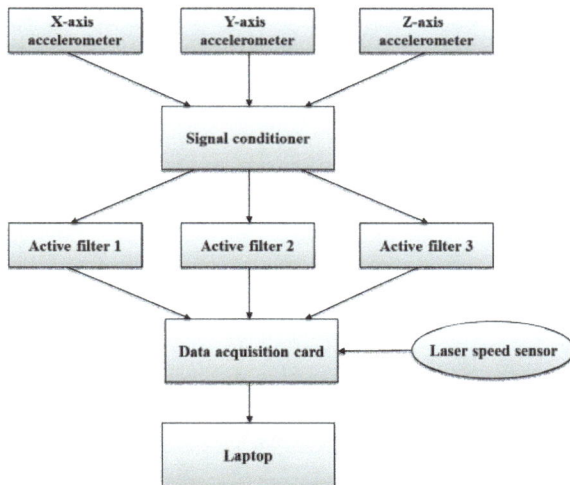

Figure 6. Schematic of the measurement set-up.

The load on the gear was applied by a magnetic brake system with the torque value of 6 lbs*in. The schematic of the measurement set-up is shown in Fig. 6.

Vibrations from three channels for axial, horizontal and vertical directions were recorded using 3 Endevco 7251A-100 accelerometers installed on the gearbox case (Fig. 4). Signal conditioning was performed using an Endevco 133 signal conditioner and analogue active anti-aliasing filters Kemo PocketMaster 1600. The cut-off frequency of the filters was set at 10 kHz. The averaged speed estimation was obtained using a laser speed sensor and reflective tape, attached to the driven wheel of the belt transmission (Fig. 3). All signals were recorded using National Instruments' data acquisition card NI DAQ-6062E at 25kHz sampling frequency.

The whole data set is represented by 7 records of approximately 3 minutes duration each with a 10-minute interval between them for damaged and undamaged gears.

The length of the signal realization for the residual signal estimation by the TSA was selected 20s according to the above-mentioned dependency of the ratio between the averaged TSA signal variance and the variance of the angular re-sampled vibration signal versus the number of averages. Therefore, each record represented by 9-10 realizations, and the whole data set of 7 records will contain 64 realizations.

4. THE DATA PROCESSING AND RESULTS

A channel selection for processing of the gear vibrations was performed according to the analysis of the standard deviation of the residual signal for undamaged and damaged gears. In Figs. 7 (a-c), the standard deviations of the TSA signals are shown in blue for the case of no damage and in red for the case when a gear tooth is damaged.

For all directions of accelerometers, values of Fisher criterion (Webb, 1999) were calculated in order to estimate the separation of standard deviation values (Table 1). According to results of this analysis, the vertical channel has been selected for further processing as the one providing the best difference of the standard deviations for the residual signals from undamaged and damaged gears.

The estimated speed required for angular re-sampling was obtained directly from the raw vibration signal (Combet & Gelman, 2007), taking the averaged speed estimation from the laser speed sensor as the input.

As it was mentioned above, the whole set of data consists of 64 realizations of data representing gears with tooth damage and 64 realizations of data representing gears without tooth damage. The training data for the damage detection were selected as follows:

- Class "damaged" was presented by 32 diagnostic features for damaged tooth 7 of the gear in 32 gear vibration signals, i.e. every other realization of the total amount of 64 realizations.

- Class "undamaged" was presented by all 18 teeth of 32 realizations for the undamaged gear and remaining 17 teeth (all, except the tooth number 7) of 32 realizations for the damaged gear (in total, 1120 diagnostic features).

	Axial direction	Horizontal direction	Vertical direction
Fisher criterion	1.0	1.4	7.9

Table 1. Values of the Fisher criterion for different directions of accelerometers.

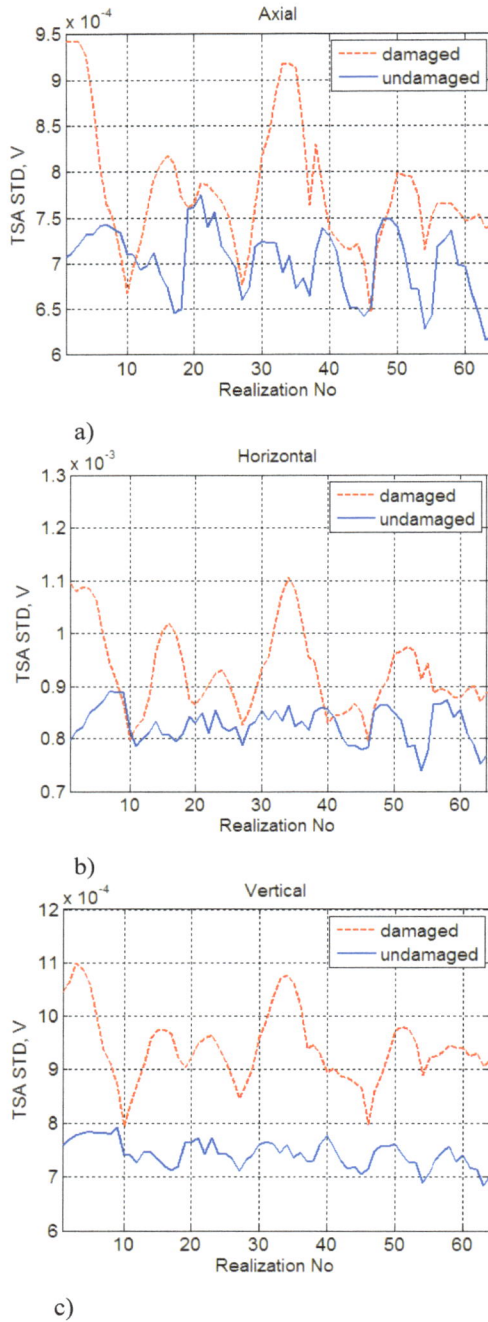

Figure 7. The standard deviations of the TSA signals for axial, horizontal and vertical directions for gear with and without tooth damage.

The averaged error probability of the correct detection was calculated for training data using the kNN classifier (Webb, 1999) for k=5:

$$P_{errAvg} = 0.5 \cdot \left(P_{errDam} + P_{errUndam} \right),$$

where P_{errDam} is the error probability for the damaged conditions, $P_{errUndam}$ is the error probability for the undamaged conditions. For the case of using the classical residual technology, the averaged error probability of correct detection is 0.14; for the case of using the novel residual technology, this probability is 0.047. Therefore, use of the novel residual technology improves separation between diagnostic features for damaged and undamaged conditions of the gear and provides almost 3 times decrement in the averaged error probability of the correct detection.

Observing the unimodal shape of diagnostic features distributions, the likelihood ratio was obtained using the Gaussian models of the data for classes "undamaged" and "damaged". To build the models, the corresponding values of mean and variance were estimated for each class. Gaussian models for classes and the resulting logarithm of the likelihood ratio are shown in Fig. 8, 9 for the classical and novel residual technologies respectively.

The likelihood ratio is estimated using the selected damaged and undamaged training data and finally, the testing data are processed. The decision making procedure used is based on the accumulated likelihood ratio which is given by:

$$\Lambda_N = \log \frac{p\left(x_1,...,x_N \mid \omega_{dam}\right)}{p\left(x_1,...,x_N \mid \omega_{undam}\right)} =$$

$$= \log \frac{\prod_i p\left(x_i \mid \omega_{dam}\right)}{\prod_i p\left(x_i \mid \omega_{undam}\right)} = \sum_{i=1}^{N_{acc}} \log \frac{p\left(x_i \mid \omega_{dam}\right)}{p\left(x_i \mid \omega_{undam}\right)},$$

where $p\left(x_i \mid \omega_{dam}\right)$ and $p\left(x_i \mid \omega_{undam}\right)$ are the conditional probability density functions for the two diagnostic classes ω_{dam} for data from damaged conditions and ω_{undam} for data from undamaged conditions; i=1,...,N_{acc}, are diagnostic features estimated on the selected sequence of realizations. N_{acc} is the number of accumulations.

The decision making rule for damage diagnosis using the sequence of realizations is:

$$\Lambda_N \geq thr_b,$$

where thr_b is the threshold for the accumulated likelihood ratio.

The diagnosis test was performed using the test data represented by another half (32 realizations) of the whole data set (64 realizations).

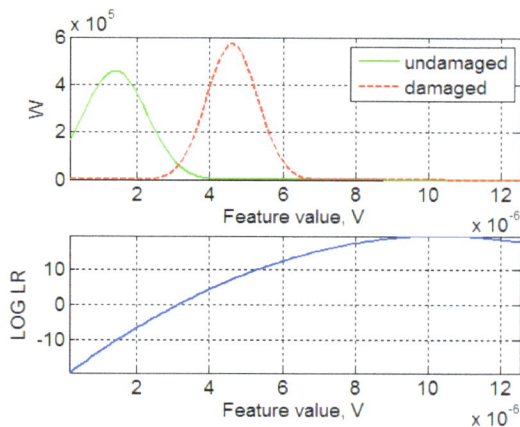

Figure 8. The Gaussian models of the probability density function W for classes "undamaged" and "damaged" (top), and logarithm of likelihood ratio (bottom) for training data based on the classical residual technology.

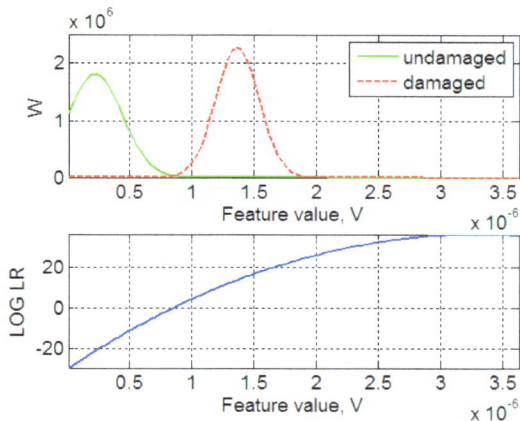

Figure 9. The Gaussian models of the probability density function W for classes "undamaged" and "damaged" (top), and logarithm of likelihood ratio (bottom) for training data based on the novel residual technology.

Selection of parameters for the test was performed as follows. The initial value of the threshold thr_b was selected lying approximately between histograms of Gaussian models of data with and without damage. The final values of thr_b and N_{acc} were obtained by optimization of the damage diagnosis procedure using the minimum of the total error probability as the optimization criterion.

For both considered cases with using the classical and the novel residual technology, the errorless diagnosis was achieved, i.e. the partly missing (chipped) tooth was diagnosed without missed detections and false alarms.

It was found that the use of the novel residual technology provides 2.3 times decrement in the diagnosis time in comparison with the classical residual technology.

5. CONCLUSIONS

• The novel residual technology for gear damage /diagnosis was successfully applied for the first time to the detection/diagnosis of the partly missing tooth.

• The gear fault: partly missing (chipped) tooth has been continuously diagnosed throughout the whole test duration without false alarms and missed detections.

• The use of the novel residual technology in comparison to the classical residual technology provides better separation of diagnostic features (i.e. the total averaged error of detection reduced 3 times for the novel residual technology) and faster damage diagnosis (i.e. diagnosis time is 2.3 times less for the novel residual technology).

ACKNOWLEDGEMENT

The authors would like to acknowledge the contribution of the Cranfield IVHM Centre experimental facilities and resource in the successful completion of this work.

REFERENCES

McFadden, P. D. (1986). Detecting fatigue cracks in gears by amplitude and phase demodulation of the meshing vibration. *Journal of Vibration, Acoustics, Stress, and Reliability in Design*, vol. 108, pp. 165-170.

Gelman, L., Zimroz, R., Birkel, J., Leigh-Firbank, H., Simms, D., Waterland, B., & Whitehurst, G. (2005). Adaptive vibration condition monitoring technology for local tooth damage in gearboxes. *Insight Int. J. Non-Destructive Testing and Condition Monitoring*, vol. 47(8), pp. 461–464.

Combet, F., Gelman, L., Anuzis, P., & Slater, R. (2009). Vibration detection of local gear damage by advanced demodulation and residual techniques. *Proc. IMechE*, vol. 223 *Part G: J. Aerospace Engineering*, pp.507-514.

Randall, R. B. (1982). A new method of modeling gear faults. *ASME Journal of Mechanical Design*, vol. 104, pp. 259–267.

Wang, W. Q., Ismail, F., & Golnaraghi, M. F. (2001). Assessment of gear damage monitoring techniques using vibration measurements. *Mechanical Systems and Signal Processing*, vol. 15(5), pp. 905–922.

Brie, D., Tomczak, M., Oehlmann, H., & Richard, A. (1997). Gear crack detection by adaptive amplitude and phase modulation. *Mechanical Systems and Signal Processing*, vol. 11(1), pp. 149–167.

Lee, S. K., & White, P. R. (1998). The enhancement of impulsive noise and vibration signals for fault detection in rotating and reciprocating machinery. *Journal of Sound and Vibration*, vol. 217(3), pp. 485–505.

Combet, F., & Gelman, L. (2009). Optimal filtering of gear signals for early damage detection based on the spectral kurtosis. *Mechanical Systems and Signal Processing*, vol. 23(3), pp. 652–668.

Wang, W., & Wong, A. K. (2002). Autoregressive model-based gear fault diagnosis. *ASME Journal of Vibration and Acoustics*, vol. 124, pp. 172–179.

Martin, N., Jaussaud, P., & Combet, F. (2004). Close shock detection using time-frequency Prony modeling. *Mechanical Systems and Signal Processing*, vol. 18(2), pp. 235–261.

Lee, J. Y., & Nandi, A. K. (2000). Extraction of impacting signals using blind deconvolution. *Journal of Sound and Vibration*, vol. 232(5), pp. 945–962.

Endo, H., & Randall, R. B. (2007). Enhancement of autoregressive model based gear tooth fault detection technique by the use of minimum entropy deconvolution filter. *Mechanical Systems and Signal Processing*, vol. 21(2), pp. 906–919.

Wang, W. J., & McFadden, P. D. (1993). Early detection of gear failure by vibration analysis—I. Calculation of the time-frequency distribution. *Mechanical Systems and Signal Processing*, vol. 7(3), pp. 193–203.

Forrester, B. D. (1996). *Advanced Vibration Analysis Techniques for Fault Detection and Diagnosis in Geared Transmission Systems*. Ph. D. Dissertation. Swinburne University of Technology, Melbourne, Australia.

Choy, F. K., Polyshchuk, V., Zakrajsek, J. J., Handschuh, R. F., & Townsend, D. P. (1996). Analysis of the effects of surface pitting and wear on the vibration of a gear transmission system. *Tribology International*, vol. 29(1), pp. 77–83.

Wang, W. J. & McFadden, P. D. (1996). Application of wavelets to gearbox vibration signals for fault detection. *Journal of Sound and Vibration*, vol. 192(5), pp. 927-939.

Loutridis, S. J. (2006). Instantaneous energy density as a feature for gear fault detection. *Mechanical Systems and Signal Processing*, vol. 20(5), pp. 1239–1253.

Halima, E. B., Shoukat Choudhuryb, M. A. A., Shaha, S. L., & Zuoc, M. J. (2008). Time domain averaging across all scales: A novel method for detection of gearbox faults. *Mechanical Systems and Signal Processing*, vol.22(2), pp.261–278.

Dalpiaz, G., Rivola, A., & Rubini, R. (2000). Effectiveness and sensitivity of vibration processing techniques for local fault detection in gears. *Mechanical Systems and Signal Processing*, vol. 14(3), pp. 387–412.

Lin, J., & Zuo, M. J. (2003). Gearbox fault diagnosis using adaptive wavelet filter. *Mechanical Systems and Signal Processing*, vol. 17(6), pp. 1259–1269.

Stewart, R. M. (1977). Some useful data analysis techniques for gearbox diagnostics. *Institute of Sound and Vibration Research*, Paper MHM/R/10/77.

McFadden, P. D. (1987). Examination of a technique for the early detection of failure in gears by signal processing of the time domain average of the meshing vibration. *Mechanical Systems and Signal Processing*, vol. 1(2), pp. 173–183.

Combet, F., & Gelman, L. (2007). An automated methodology for performing time-synchronous averaging of a gearbox signal without speed sensor. *Mechanical Systems and Signal Processing*, vol. 21(6), pp. 2590-2606.

Webb, A. (1999). *Statistical pattern recognition*. London: Arnold.

Remaining Useful Lifetime (RUL): Probabilistic Predictive Model

Ephraim Suhir

University of California, Santa Cruz, CA, 95064-1077, USA
suhire@aol.com

ABSTRACT

Reliability evaluations and assurances cannot be delayed until the device (system) is fabricated and put into operation. Reliability of an electronic product should be conceived at the early stages of its design; implemented during manufacturing; evaluated (considering customer requirements and the existing specifications), by electrical, optical and mechanical measurements and testing; checked (screened) during fabrication; and, if necessary and appropriate, maintained in the field during the product's operation. Prognostics and health monitoring (PHM) effort can be of significant help, especially at the last, operational stage, of the product use. Accordingly, a simple and physically meaningful probabilistic predictive model is suggested for the evaluation of the remaining useful lifetime (RUL) of an electronic device (system) after an appreciable deviation from its normal operation conditions has been detected, and the corresponding increase in the failure rate and the change in the configuration of the wear-out portion of the bathtub curve has been assessed. The general concepts are illustrated by a numerical example. The model can be employed, along with other PHM forecasting and interfering tools and means, to evaluate and to maintain the high level of the reliability (probability of non-failure) of a device (system) at the operation stage of its lifetime.

1. INTRODUCTION

Reliability evaluation and assurance cannot be delayed until the device is fabricated and launched into operation, although it is sometime the case in many current practices. Reliability of an electronic product should be conceived at the early stages of its design; implemented during manufacturing; evaluated (considering customer requirements and the existing specifications), by electrical, optical and mechanical measurements and testing; checked (screened) during manufacturing (fabrication); and, if necessary and appropriate, maintained in the field during the product's operation (see. e.g., Suhir, 1997). The prognostics-and-health-monitoring (PHM) concepts and techniques (see, e.g., Vichare and Pecht, 2006; Kirkland, Pombo, Nelson and

Berghout, 2004) are viewed today as an important part of electronic product reliability assurance at the last, operational, stage of the product's life, when there is a need and a possibility to maintain the product's high operational reliability in the field. As is known, PHM is based on a continuous monitoring of the products behavior in the field and is aimed at the prediction of the future reliability of the product from the detected and assessed deviation of its performance (because of aging, degradation, elevated loading condition, extraordinary and harsh environment, etc.) from the normal (specified) performance. The ability to predict the remaining useful lifetime (RUL), after a certain malfunction is detected or anticipated, is one of the most crucial PHM objectives.

Accordingly, in the analysis that follows, we suggest a formalism for the assessment of the RUL of a device (system) from 1) the given steady-state failure rate (FR), 2) agreed-upon ultimate failure rate (beyond which the further use of the device or a system is deemed undesirable), 3) the increase in the FR at the wear-out portion of the bathtub curve and 4) the detected or anticipated small "jump" in the FR that determines the beginning of the actual wear-out stage. This jump could be determined in many ways, depending on the physical nature of the addressed degradation (aging) process, typical loads (usually thermally induced), available PHM equipment and its trustworthiness, etc. Having in mind the inevitable uncertainties in the magnitude of such a "jump", we use, in addition to the "deterministic" formalism, also a formalism based on the probabilistic design-for-reliability (PDfR) concept (see Suhir, 2010 and Suhir, Mahajan, Lucero and Bechou, 2012).

2. ANALYSIS

2.1. Remaining Useful Lifetime (RUL)

Let the wear-out portion $\lambda(t)$ of the original (specified) bathtub curve is configured as shown in Fig.1 (solid line). The commencement of this portion is defined by the moment of time when a (typically insignificant) malfunction in the device performance is detected, and the PHM instrumentation and algorithms predict a small "jump" $\Delta\lambda$ in the bathtub curve. The new (corrected) configuration $\tilde{\lambda}(t)$ of the wear out portion of the bathtub curve is shown as a broken line in Fig.1. The "healthy" and "damaged" configurations of this portion of the bathtub curve can be approximated as

$$\lambda(t) = \lambda_0 t^n, \qquad \tilde{\lambda}(t) = \tilde{\lambda}_0 t^{\tilde{n}} \qquad (1)$$

respectively. The following notation is used:

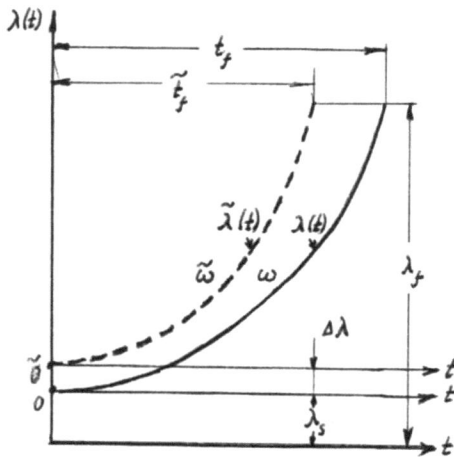

Fig.1. Specified/"healthy" (solid line) and the deviated/"damaged" (broken line) configurations of the wear out portion of the bathtub curve.

$$\lambda_0 = \frac{\lambda_f - \lambda_s}{t_f^n},$$

$$n = \frac{\lambda_f - \lambda_s}{\omega} t_f - 1 = \frac{1 - \beta}{\beta},$$

$$\tilde{\lambda}_0 = \frac{\lambda_f - \lambda_s - \Delta\lambda}{\tilde{t}_f^{\tilde{n}}},$$

$$\tilde{n} = \frac{\lambda_f - \lambda_s - \Delta\lambda}{\tilde{\omega}} \tilde{t}_f - 1 = \frac{1 - \tilde{\beta}}{\tilde{\beta}}, \qquad (2)$$

λ_f is the specified maximum acceptable failure rate beyond which the further use of the device or a system is deemed unfeasible, λ_s is the steady-state failure rate, t_f is the specified duration of the wear-out period of the bath-tub curve, ω is the (shaded) area under the curve $\lambda(t)$, $\beta = \dfrac{1}{n+1}$ is the area coefficient (reflecting the degree of deviation of the area in question from a rectangular), $\tilde{t}_f$ is the RUL, $\tilde{\omega}$ is the (shaded) area under

the curve $\tilde{\lambda}(t)$ and $\tilde{\beta} = \dfrac{1}{\tilde{n}+1}$ is the corresponding area coefficient. The first and the third formulas in (2) are, in effect, conditions that reflect the requirement that the device is not supposed to be operated beyond the λ_f FR level.

In an approximate analysis, in order to assess the sensitivity of the $\tilde{t}_f$ value to the change in the major factors affecting the RUL after an appreciable deviation from the specified ("standard", "anticipated") wear out portion of the bathtub curve has been diagnosed and assessed, we assume $\lambda_0 \approx \tilde{\lambda}_0$ and $\beta \approx \tilde{\beta}$. Then the first and the third formulas in (2) result in the following formula for the relative RUL:

$$\tau_u = \frac{\tilde{t}_f}{t_f} \approx (1 - \xi)^{\frac{\beta}{1-\beta}}, \qquad (3)$$

where

$$\xi = \frac{\Delta\lambda}{\lambda_f - \lambda_s} \qquad (4)$$

is the ratio of the predicted increase in the failure rate to the difference between the ultimate failure rate and the steady state failure rate.

The following conclusions can be made based on the formula (3):

- The relative RUL τ_u changes from one to zero, when the predicted damage ratio ξ changes from zero to one.

- The relative RUL τ_u changes from $1 - \xi$ to zero, when the "fullness" parameter β changes from 0.5 to one. The condition $\dfrac{d\tau_u}{d\beta} = 0$ indicates that the $\tau_u = 1 - \xi$ value is the maximum possible value of the τ_u ratio. This value takes place for zero ξ values, i.e., for zero "damages" $\Delta\lambda$.

- The relative RUL increases with an increase in the parameter β (decrease in the exponent n), i.e., with an increase in the "fullness" of the wear-out portion of the bathtub curve. In other words, it is advisable that the wear-out portion of the bathtub curve "concentrates" at the end of the bathtub diagram;

- The relative RUL increases with a decrease in the predicted "disturbance" $\Delta\lambda$ and the increase in the difference between the ultimate (specified) failure rate λ_f and the steady-state failure rate λ_s.

- Lower steady-state failure rates λ_s result in larger RUL.

These intuitively more or less obvious conclusions are quantified by the formulas (3) and (4).

Let us show, as an illustration, how the possible (expected) "damage" $\Delta\lambda$, if attributed to the change in temperature, can be assessed from the detected appreciable increase ΔT in temperature. If one chooses the Boltzmann-Arrhenius law

$$\overline{\tau} = \tau_0 \exp\left(\frac{U}{kT}\right) \qquad (5)$$

to determine the mean-time-to-failure $\overline{\tau}$ and the corresponding failure rate $\lambda = \dfrac{1}{\overline{\tau}}$ for the given absolute temperature T, then the derivative $\dfrac{d\lambda}{dT}$ can be found as

$$\frac{d\lambda}{dT} = -\frac{U}{\tau_0 kT^2} \exp\left(-\frac{U}{kT}\right) \qquad (6)$$

Here U is the activation energy, k is Boltzmann's constant, λ_s is the steady-state failure rate, and T_s is the steady-state temperature (prior to the detected temperature increase). Replacing the differentials in the above formula with finite differences, we have:

$$\Delta\lambda = -\frac{U}{\tau_0 kT^2} \exp\left(-\frac{U}{kT}\right)\Delta T = -\frac{U\lambda_s}{kT_s^2}\Delta T . \quad (7)$$

The formula (7) indicates particularly that the "damage" will be lower for lower steady-state failure rates and for higher steady-state operation temperatures.

2.2. Probabilistic Approach

The probabilistic design for reliability (PDfR) approach enables one to account for the random nature of the relative RUL τ_u as a non-random function of the random variable ξ.

Let, for the sake of simplicity, assume that this variable is distributed in accordance with the Rayleigh's law:

$$f_\xi(\xi) = \frac{\xi}{\xi_0^2} \exp\left(-\frac{\xi^2}{2\xi_0^2}\right) , \qquad (8)$$

where ξ_0 is the most likely (deterministic) value of the variable ξ. The physical justification for the taken assumption that the random variable ξ is distributed in accordance with the Rayleigh's law is that a random variable of time should always be positive, that low ξ values are more likely than high ξ values, that the zero value of the variable ξ should be equal to zero, and the probabilities of its high values should be very small and

should decrease with an increase in the ξ value. Weibull distribution and normal distributions with significant mean-to-standard-deviation ratios could be assumed in practical applications as more flexible and "richer" two-parametric distributions for the variable ξ.

The probability density function of the variable τ_u can be found as

$$f_\tau(\tau_u) =$$

$$= \frac{1-\beta}{\beta\xi_0^2} \tau_u^{\frac{1}{\beta}-2}\left(1 - \tau_u^{\frac{1}{\beta}-1}\right)\exp\left[-\frac{\left(1 - \tau_u^{\frac{1}{\beta}-1}\right)^2}{2\xi_0^2}\right] , (9)$$

and the probability distribution function is

$$F_\tau(\tau_u) = \exp\left[-\frac{\left(1 - \tau_u^{\frac{1}{\beta}-1}\right)^2}{2\xi_0^2}\right] \qquad (10)$$

This function determines the probability that the random RUL ratio (3) will not exceed a certain τ_u level. It is clear that one wishes to assure that this probability is as low as possible.

When the PDfR approach is used, there is of interest to compare the probabilities of non-failure of the "healthy" and the "damaged" device (system). Assuming, again, for the sake of simplicity, that the exponential law of reliability is applicable, one can use the following formula to evaluate the probability of non-failure:

$$P = \exp(-\lambda t). \qquad (11)$$

The time runs faster, by the factor of $1/\tau$, in the case of a "damaged" device, so that the formula (11) yields:

$$\widetilde{P} = P^{1/\tau} \qquad (12)$$

where the τ ratio is defined by the formula (3).

3. NUMERICAL EXAMPLE

Input data:

Steady-state failure rate: $\lambda_s = 2.5 x 10^{-5} 1/hr$;

Ultimate failure rate: $\lambda_u = 7.5 x 10^{-5} 1/hr$;

Area ("fullness") parameter of the wear-out portion of the bathtub curve: $\beta = 0.75$

Calculated data:

The calculated data for the relative RUL τ_u ratios for different changes $\Delta\lambda$ in the initial values are shown in Table 1. These data indicate that the RUL of the "damaged" device (system) rapidly decreases with an increase in the level of the initial "damage". This "damage" is defined in our analysis as the relative deviation of the failure rate at the wear-out portion of the bathtub curve.

$\frac{\Delta\lambda}{1/hr}$	0.10 E-5	0.25 E-5	0.75 E-5	1.0 E-5	1.5 E-5	2.5 E-5
ξ	0.02	0.05	0.15	0.20	0.30	0.5
τ_u	0.9412	0.8574	0.6141	0.6672	0.3430	0.125

Table 1. Calculated relative remaining useful life (RUL) data vs. change in the failure rate

ξ_0	0.02	0.04	0.06	0.08	0.10
τ_u	x	x	x	x	x
0.2	0	0	0	0	0.0002
0.4	0	0	0	0.0045	0.0313
0.6	0	0.0005	0.0332	0.1472	0.2935
0.8	0.001625	0.2008	04899	0.6694	0.7734
0.9	0.2257	0.6892	0.8475	0.9112	0.9422
1.0	1.0	1.0	1.0	1.0	1.0

Table 2. Calculated probabilities of the situation that the actual remaining-useful-life (RUL) ratio for the "damaged" and "healthy" devices is below the remaining relative useful life level

The calculated probabilities $F(\tau_u)$ of the situation that the actual ratio of the RUL of the "damaged" and "healthy" devices remains below the τ_u level is calculated for different τ_u values and different most likely values ξ_0 of the relative damage ξ are shown in Table 2. The calculated data indicate that the sought probability is small indeed (which is certainly a desirable situation) for low τ_u and low ξ_0 values, when the most likely damage ξ_0 is small and/or when the RUL of the damaged device is short compared to the RUL of the "healthy" device. This means that the probability $P(\tau_u) = 1 - F(\tau_u)$ that the actual RUL ratio will exceed the τ_u value indicated in the left column of the Table 2 is high.

τ_u	0.2	0.4	0.6	0.8	0.9	1.0
P	x	x	x	x	x	x
E-2	E-10	E-5	4.642 E-4	3.162 E-3	5.995 E-3	E-2
E-3	E-15	3.162 E-8	E-5	1.778 E-4	4.642 E-4	E-3
E-4	E-20	E-10	2.154 E-7	E-5	3.594 E-5	E-4
E-5	E-25	3.16 E-13	4.642 E-9	5.623 E-7	2.783 E-6	E-5

Table 3. Calculated decrease in the probability $\widetilde{P}$ of non-failure, at the given level of the failure rate, of the "damaged" device as compared to the probability P of non-failure of the "healthy" one.

As to the data at the lower right corner of the Table 3, they indicate that it is very likely that the actual RUL of the damaged device will be considerably lower than the desirably high τ_u value. The Table 3, obtained on the basis of the formula (12), shows the decrease in the probability $\widetilde{P}$ of non-failure, at the given level of the failure rate, of the "damaged" device as compared to the probability P of non-failure of the "healthy" one. The probabilities $\widetilde{Q} = 1 - \widetilde{P}$ of failure are high for low τ_u ratios, especially if the specified probabilities P of non-failure for a "healthy" device are low.

4. CONCLUSION

Simple, easy-to-use and physically meaningful predictive formalisms are developed for the evaluation of the remaining useful life (RUL) after an appreciable deviation from the normal operation conditions has been detected and the change in the wear-out portion of the bathtub curve has been predicted. Both deterministic and probabilistic approaches are considered. The models can be used, in addition to other PHM forecasting means, in the analysis and design of various PHM systems, including the PDfR.

REFERENCES

Kirkland, L.V., Pombo, T, Nelson, K., Berghout, F. (2004). Avionics Health Management: Searching for the Prognostics Grail, *Proceedings of IEEE Aerospace Conference,* Vol. 5. Big Sky, MO

Suhir, E. (1997). *Applied Probability for Engineers and Scientists*, McGraw-Hill

Suhir, E. (2010). Probabilistic Design for Reliability, ChipScale Reviews, vol.14, No.6.

Suhir, E., Mahajan, R., Lucero, A. and Bechou, L. (2012). Probabilistic Design-for-Reliability Concept and Novel Approach to Qualification Testing of Aerospace Electronic Products, *Proceedings of IEEE Aerospace Conference,* Vol. 5. Big Sky, MO (to be presented and published)

Vichare, N., Pecht, M. (2006). Prognostics and Health Management of Electronics, IEEE Transactions on Components and Packaging Technologies, Vol. 29, No. 1

Ephraim Suhir is on the *faculty* of the University of California, Santa Cruz, CA, Electrical Engineering Department. He is also Visiting Professor, Mechanical Engineering Department, University of Maryland, College Park, MD and Department of Electronics Materials, Technical University, Vienna, Austria. He is *Fellow* of the Institute of Electrical and Electronics Engineers (IEEE), the American Physical Society (APS), the American Society of Mechanical Engineers (ASME), the Institute of Physics (IoP), UK, and the Society of Plastics Engineers (SPE). Dr. Suhir has been elected as Foreign Full Member (*Academician*) of the National Academy of Engineering and Technological Sciences, Ukraine. He is on the US Department of State roster as *Fulbright Scholar* in information and telecommunication technologies.

Dr. Suhir is a co-founder of the ASME Journal of Electronic Packaging and served as its Technical Editor (Editor-in-Chief) for eight years (1993-2001). He holds *22 US patents* and has authored about *300 technical publications* (papers, book chapters, books), including monographs "Probabilistic Methods in Ship Structural Analysis", Nikolayev Institute of Naval Architecture, Nikolayev, Ukraine, 1973 (in Russian), "Structural Analysis in Microelectronics and Fiber Optics", Van-Nostrand, 1991, and "Applied Probability for Engineers and Scientists", McGraw-Hill, 1997. Dr. Suhir is editor of the Springer book series on physics, mechanics and packaging of microelectronic and photonic systems. He is

Distinguished Lecturer of the IEEE CPMT (Components, Packaging and Manufacturing Technology) Society, serves on several Technical Committees of this Society and is *Associate Editor* of the IEEE CPMT Transactions on Advanced Packaging.

Dr. Suhir received many distinguished service and professional awards, including: *2004 ASME Worcester Read Warner Medal* for outstanding contributions to the permanent literature of engineering through a series of papers in Mechanical, Microelectronic, and Optoelectronic Engineering, and is the third Russian American (after Igor Sikorsky and Stephen Timoshenko) who received this prestigious award; *2001 IMAPS John A. Wagnon Technical Achievement Award* for outstanding contributions to the technical knowledge of the microelectronics, optoelectronics, and packaging industry; *2000 IEEE-CPMT Outstanding Sustained Technical Contribution Award* for outstanding, sustained and continuing contributions to the technologies in fields encompassed by the CPMT Society; *2000 SPE International Engineering/Technology (Fred O. Conley) Award* for outstanding pioneering and continuing contributions to plastics engineering; *1999 ASME and Pi-Tau-Sigma Charles Russ Richards Memorial Award* for outstanding contributions to mechanical engineering, and *1996 Bell Laboratories Distinguished Member of Technical Staff Award* for developing engineering mechanics methods for predicting the reliability, performance, and mechanical behavior of complex structures used in manufacturing AT&T and Lucent Technologies products.

Author Index

Author Guidelines

The International Journal of Prognostics and Health Management (IJPHM) publishes scientific papers dealing with all aspects of prognostics, diagnostics, and system health management of complex engineered systems. High quality articles focused on assessing the current status and predicting the future condition of an engineered component and/or system of components. Such articles may come from a variety of disciplines, including electrical, electronics, mechanical, civil, and chemical engineering, computer and materials science, reliability, test and measurement, artificial intelligence, physics, and economics.

Copyright

The Prognostic and Health Management Society advocates open-access to scientific data and uses a Creative Commons license for publishing and distributing any papers. A Creative Commons license does not relinquish the author's copyright; rather it allows them to share some of their rights with any member of the public under certain conditions whilst enjoying full legal protection. By submitting an article to the International Conference of the Prognostics and Health Management Society, the authors agree to be bound by the associated terms and conditions including the following: As the author, you retain the copyright to your Work. By submitting your Work, you are granting anybody the right to copy, distribute and transmit your Work and to adapt your Work with proper attribution under the terms of the Creative Commons Attribution 3.0 United States license. You assign rights to the Prognostics and Health Management Society to publish and disseminate your Work through electronic and print media if it is accepted for publication. A license note citing the Creative Commons Attribution 3.0 United States License, as shown below, needs to be placed in the footnote on the first page of the article.

First Author et al. This is an open-access article distributed under the terms of the Creative Commons Attribution 3.0 United States License, which permits unrestricted use, distribution, and reproduction in any medium, provided the original author and source are credited.

Ethics

Contributions to IJPHM must report original research and will be subjected to review by referees at the discretion of the Editor. IJPHM considers only manuscripts that have not been published elsewhere (including at conferences), and that are not under consideration for publication or in press elsewhere. Moreover, it is the responsibility of the author to ensure that any data or information submitted complies with the export-control regulations of the author's home country (e.g., International Traffic in Arms Regulations (ITAR) in the United States). IJPHM honors code of conduct provided by the Committee of Publication Ethics (COPE). More details on IJPHM policies and publication ethics can be found online.

Submission Types

IJPHM publishes full-length regular papers, technical briefs, communications, and survey papers.

Full-Length Regular Papers should describe new and carefully confirmed findings, and experimental procedures and results should be given in detail sufficient for others to replicate the work. A full paper should be long enough to describe and interpret the work clearly, placing it in the context of other research.

Technical Briefs usually describe a single result, experiment, or technique of general interest for which a short treatment is appropriate. A short paper should be long enough to describe experimental procedures and clearly, and interpret the results in the context of other research.

Communications are a separate class of short manuscripts that are subject to an expedited review process. Appropriate items include (but are not limited to) rebuttals and/or counterexamples of previously published papers. A short communication is suitable for recording the results of complete small investigations or giving details of new models or hypotheses, innovative methods, techniques or apparatus. The style of main sections need not conform to that of full-length papers. Short communications are 2 to 4 printed pages in length. The Editors will review these submissions internally, and request outside review when appropriate.

Survey Papers covering emerging research topics in PHM are also published, and unsolicited manuscripts of a tutorial or review nature are welcome. However, prospective authors of survey papers should contact in advance the Editor-in-Chief in order to assess the possible interest of the topic to IJPHM. Papers describing specific current applications are encouraged, provided that the designs represent the best current practice, detailed characteristics and performance are included, and they are of general interest.

Prospective authors should note that for any type of IJPHM content, poorly documented papers using "proprietary" techniques will be rejected. Moreover, excessive "branding" within a paper also cause for rejection; e.g., "The team used the magical CompanyBrand™ preprocessing to prepare the data to extract the amazing CompanyBrand™-proprietary features (which we can't tell you about)." Papers should present techniques and results clearly and objectively.

Although bound editions will be available for purchase, IJPHM is fundamentally an online journal. As such, we are able to have a very fast turnaround time. We will acknowledge receipt of submissions within three business days, and we intend to rigorously review and return a decision to the authors in approximately 8-12 weeks. Thus, papers may be published in a very short time, allowing your research to be available to the scientific community when it is most relevant.

Option to Present Your Work at a Conference

PHM Society publications have maintained high quality standards for both its Conferences and the Journal. Highest quality conference papers are also invited to be published in the Journal. However, since 2012 IJPHM provides an option to the journal authors to present their journal paper at one of the upcoming PHM conferences.

Authors are reminded that the paper must be journal quality and adhering to the journal template. The paper will be reviewed as per journal review standards and if accepted a presentation slot will be reserved at the target conference. The paper will be published in the journal archives and linked through conference proceedings.

Benefits
- A journal publication of your high quality research work
- A peer review of your work by experts in the field
- A chance to present your work to the targeted audience
- No reworking required to publish in the Journal
- A shortened review cycle to journal publishing

Risks
- Rejected papers will not automatically be considered for the conference and may additionally miss the submission deadline.
- If re-submitted for the conference, they will be reviewed subject to conference review criteria

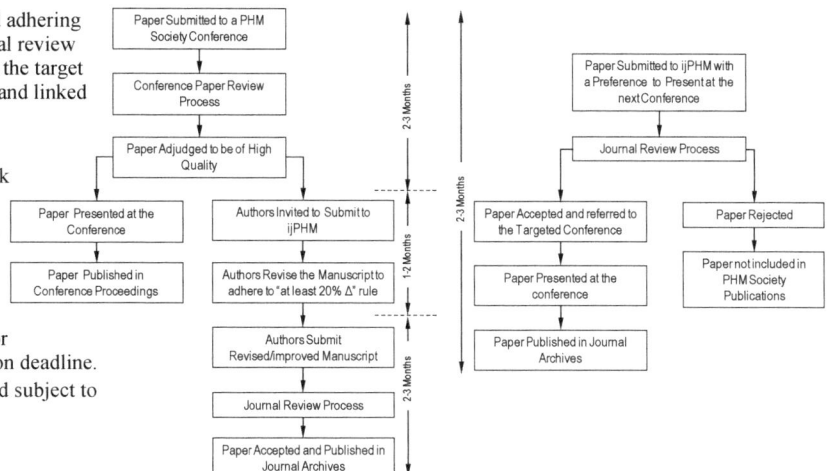

www.ingramcontent.com/pod-product-compliance
Lightning Source LLC
Chambersburg PA
CBHW041713210326
41598CB00007B/639